It's another great book from CGP...

Chemistry exams can seem daunting — especially if you're not sure what to expect.
But they're less stressful if you've done plenty of realistic practice in advance.

Happily, this book (which includes a **free** Online Edition)
is packed with exam-style questions for every topic. It even includes
two complete practice exams to fully prepare you for the real thing.

How to get your free online edition

Want to read this book on your computer or tablet?
Just go to **cgpbooks.co.uk/extras** and enter this code...

0220 6686 2723 6520

By the way, this code only works for one person. If somebody else has used
this book before you, they might have already claimed the online edition.

D1630834

CGP — still the best! ☺

Our sole aim here at CGP is to produce the highest quality books —
carefully written, immaculately presented and dangerously close to being funny.

Then we work our socks off to get them out to you
~~at the cheapest possible prices.~~

Contents

Use the tick boxes to check off the topics you've completed.

Section One — Fundamental Ideas in Chemistry

Section Two — Bonding and Structure

Section Three — Air and Water

Section Four — The Periodic Table and Metals

Section Five — Acids, Bases and Reaction Rates

Section Six — Crude Oil and Organic Chemistry

Section Seven — Energy and Equilibria

Section Eight — Electrolysis and Analysis

Practice Papers

How to get answers for the Practice Papers
Your free Online Edition of this book includes all the answers for Practice Papers 1 & 2.
(Just flick back to the previous page to find out how to get hold of your Online Edition.)

Published by CGP

Editors:
Katie Braid, Katherine Craig, Mary Falkner, Ben Fletcher,
Christopher Lindle, Sean Stayte, Charlotte Whiteley.

Contributors:
Michael Aicken, Max Fishel, Paddy Gannon, Anne Hetherington.

ISBN: 978 1 84762 449 9

With thanks to Chris Elliss, Matteo Orsini Jones and Glenn Rogers for the proofreading.
With thanks to Jamie Sinclair for the reviewing.

Groovy website: www.cgpbooks.co.uk

Jolly bits of clipart from CorelDRAW®
Printed by Elanders Ltd, Newcastle upon Tyne

Based on the classic CGP style created by Richard Parsons.

How to Use This Book

- Hold the book <u>upright</u>, approximately <u>50 cm</u> from your face, ensuring that the text looks like <u>this</u>, not ~~sᴉɥʇ~~. Alternatively, place the book on a <u>horizontal</u> surface (e.g. a table or desk) and sit adjacent to the book, at a distance which doesn't make the text too small to read.

- In case of emergency, press the two halves of the book together <u>firmly</u> in order to close.

- Before attempting to use this book, familiarise yourself with the following <u>safety information</u>:

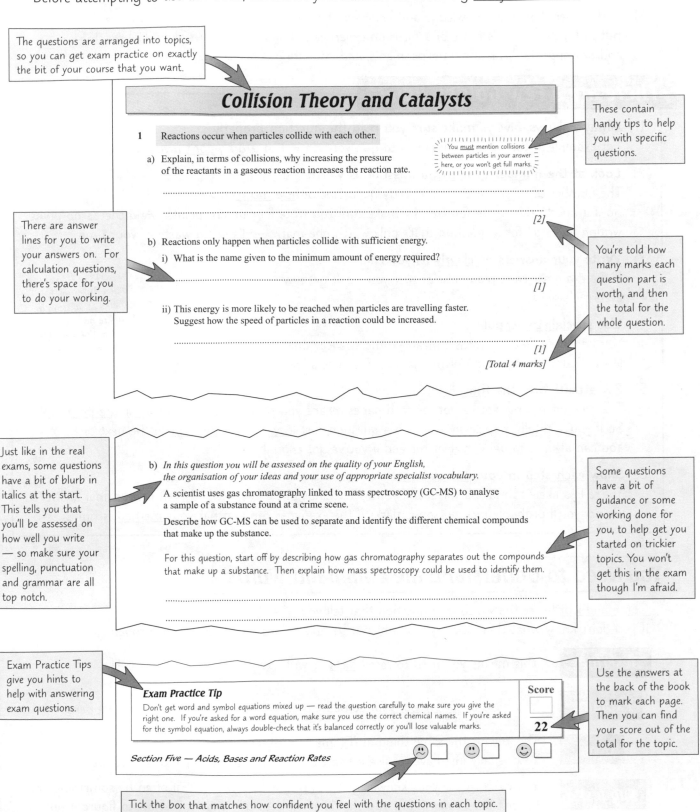

The questions are arranged into topics, so you can get exam practice on exactly the bit of your course that you want.

Collision Theory and Catalysts

1   Reactions occur when particles collide with each other.

a) Explain, in terms of collisions, why increasing the pressure of the reactants in a gaseous reaction increases the reaction rate.

....

....

[2]

b) Reactions only happen when particles collide with sufficient energy.

i) What is the name given to the minimum amount of energy required?

....

[1]

ii) This energy is more likely to be reached when particles are travelling faster. Suggest how the speed of particles in a reaction could be increased.

....

[1]

[Total 4 marks]

You <u>must</u> mention collisions between particles in your answer here, or you won't get full marks.

These contain handy tips to help you with specific questions.

There are answer lines for you to write your answers on. For calculation questions, there's space for you to do your working.

You're told how many marks each question part is worth, and then the total for the whole question.

b) *In this question you will be assessed on the quality of your English, the organisation of your ideas and your use of appropriate specialist vocabulary.*

A scientist uses gas chromatography linked to mass spectroscopy (GC-MS) to analyse a sample of a substance found at a crime scene.

Describe how GC-MS can be used to separate and identify the different chemical compounds that make up the substance.

For this question, start off by describing how gas chromatography separates out the compounds that make up a substance. Then explain how mass spectroscopy could be used to identify them.

....

....

Just like in the real exams, some questions have a bit of blurb in italics at the start. This tells you that you'll be assessed on how well you write — so make sure your spelling, punctuation and grammar are all top notch.

Some questions have a bit of guidance or some working done for you, to help get you started on trickier topics. You won't get this in the exam though I'm afraid.

Exam Practice Tips give you hints to help with answering exam questions.

Exam Practice Tip

Don't get word and symbol equations mixed up — read the question carefully to make sure you give the right one. If you're asked for a word equation, make sure you use the correct chemical names. If you're asked for the symbol equation, always double-check that it's balanced correctly or you'll lose valuable marks.

Score

22

Use the answers at the back of the book to mark each page. Then you can find your score out of the total for the topic.

Section Five — Acids, Bases and Reaction Rates

Tick the box that matches how confident you feel with the questions in each topic. This should help show you where you need to focus your revision.

Exam Tips

AQA Certificate Exam Stuff

1) You have to do two exams for the AQA Level 1/2 Certificate in Chemistry — Paper 1 and Paper 2 (ingenious).

2) Both exams are 1½ hours long, and worth 90 marks.

3) Both papers test your knowledge and understanding of Chemistry. No surprises there. But in Paper 2, there's more of a focus on experimental and investigative skills, like reading and drawing graphs, planning experiments, and evaluating conclusions.

There are a Few Golden Rules:

1) **Always, always, always make sure you read the question properly.**
For example, if the question asks you to give your answer in g/dm^3, don't give it in mol/dm^3.

2) **Look at the number of marks a question is worth.**
The number of marks gives you a pretty good clue of how much to write.
So if a question is worth four marks, make sure you write four decent points. And there's no point writing an essay for a question that's only worth one mark — it's just a waste of your time.

3) **Write your answers as clearly as you can.**
If the examiner can't read your answer you won't get any marks, even if it's right.

4) **Use specialist vocabulary.**
You know the kind of words I mean — the sciencey ones, like enthalpy change and polymerisation. Examiners love them.

Obeying these Golden Rules will help you get as many marks as you can in the exam — but they're no use if you haven't learnt the stuff in the first place. So make sure you revise well and do as many practice questions as you can.

5) **Pay attention to the time.**
The amount of time you've got for each paper means you should spend about a minute per mark. So if you're totally, hopelessly stuck on a question, just leave it and move on to the next one. You can always go back to it at the end if you've got enough time.

6) **Show each step in your calculations.**
You're less likely to make a mistake if you write things out in steps. And even if your final answer's wrong, you'll probably pick up some marks if the examiner can see that your method is right.

You Need to Understand the Command Words

Command words are the words in a question that tell you what to do.
If you don't know what they mean, you might not be able to answer the questions properly.

Describe... This means you need to recall facts or write about what something is like.

Explain... You have to give reasons for something or say why or how something happens.

Give... This means the same thing as 'Name...' or 'State...'.
You usually just have to give a short definition or an example of something.

Suggest... You need to use your knowledge to work out the answer. It'll often be something you haven't been taught, but you should be able to use what you know to figure it out.

Calculate... This means you'll have to use numbers from the question to work something out.
You'll probably have to get your calculator out.

States of Matter

1 The photograph below shows a vessel in a distillery. The walls of the vessel are solid copper.

a) Use words from the box to complete the sentences about solids.
Each word may be used once, more than once or not at all.

| weak | move | colder | hotter |
| strong | expand | heavier | dissolve |

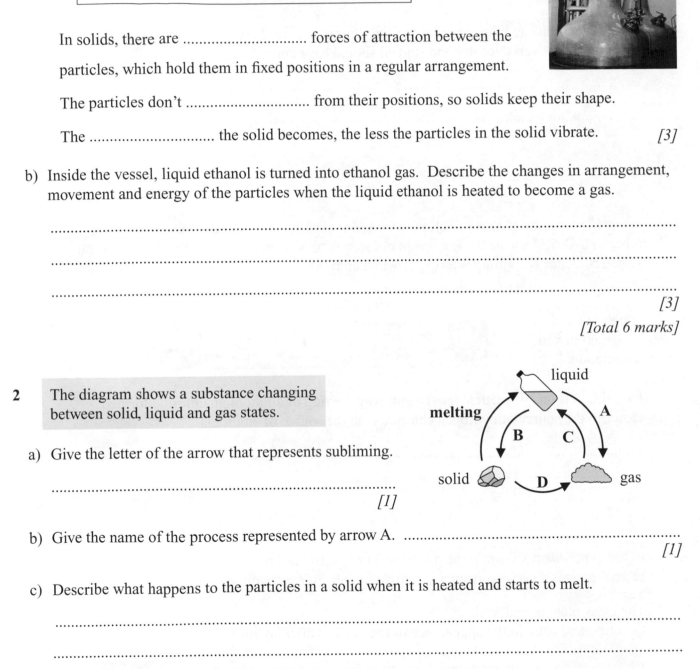

In solids, there are .... forces of attraction between the

particles, which hold them in fixed positions in a regular arrangement.

The particles don't .... from their positions, so solids keep their shape.

The .... the solid becomes, the less the particles in the solid vibrate. *[3]*

b) Inside the vessel, liquid ethanol is turned into ethanol gas. Describe the changes in arrangement, movement and energy of the particles when the liquid ethanol is heated to become a gas.

....

....

....
[3]

[Total 6 marks]

2 The diagram shows a substance changing between solid, liquid and gas states.

a) Give the letter of the arrow that represents subliming.

....
[1]

b) Give the name of the process represented by arrow A. ....
[1]

c) Describe what happens to the particles in a solid when it is heated and starts to melt.

....

....

....
[3]

[Total 5 marks]

Score:

11

Movement of Particles

1 A student placed 0.2 g of potassium manganate(VII) in a beaker of water at room temperature. The potassium manganate(VII) completely dissolved, turning the water nearby purple. Eventually all the water in the beaker was purple. The student timed how long it took for the colour to spread through the entire beaker.

a) Give the name of the process which made the colour spread through the beaker.

....
[1]

b) The student repeated the experiment using the same mass of potassium manganate(VII). Give **two** other variables that the student should have controlled.

....

....
[2]

[Total 3 marks]

2 A teacher took two pieces of cotton wool and soaked one in aqueous ammonia and the other in hydrochloric acid. The teacher placed them at opposite ends of a glass tube and sealed it. After five minutes a ring of white ammonium chloride had formed on the tube.

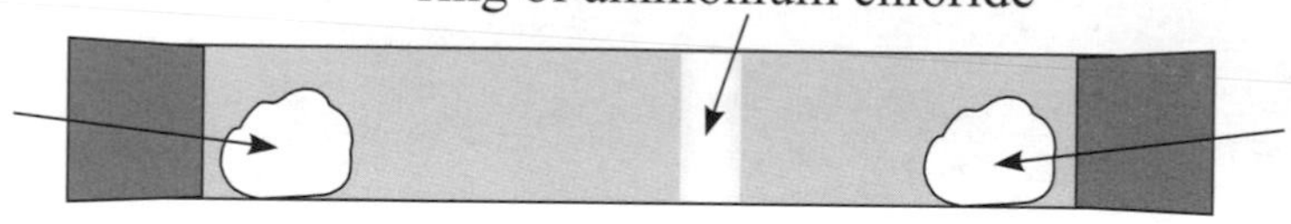

cotton wool soaked in aqueous ammonia

ring of ammonium chloride

cotton wool soaked in hydrochloric acid

Explain, in terms of particle movement, why the ring of ammonium chloride formed closer to the source of hydrogen chloride than the source of ammonia.

....

....
[Total 2 marks]

3 In the experiment shown in the diagram, a gas jar full of brown bromine gas is separated from a gas jar full of air by a glass plate.

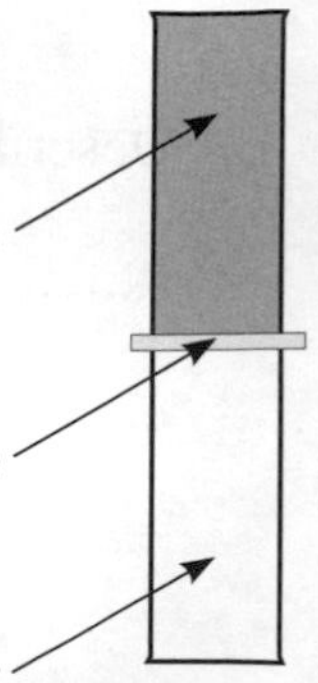

The glass plate is removed.
Describe and explain the appearance of the gas jars after an hour.

bromine

glass plate

air

....

....

....
[Total 2 marks]

Score: [ ]

7

Atoms

1 Atoms are made up of protons, neutrons and electrons.

a) Complete the table below showing the relative mass and charge of these particles.

Particle	Relative mass	Relative Charge
Proton	....	+1
Neutron	1	....
Electron	very small	....

[3]

b) A helium atom contains two protons and two electrons.

i) Explain why a helium atom has no overall charge.

....

....

[2]

ii) Draw a diagram of a helium atom. Label each type of particle on your diagram.

[3]

c) There are about 100 different *elements*.

i) What is meant by the term *element*?

....

....

[1]

ii) Complete the sentence

All atoms of a particular element contain the same number of ....

[1]

[Total 10 marks]

Score: ☐

10

Section One — Fundamental Ideas in Chemistry

Atoms and the Periodic Table

1   The periodic table displays all the known elements arranged into groups.

a) Complete the sentence.

In the periodic table each .... represents a group of elements.

[1]

b) Where in the periodic table are the non-metallic elements located?

....

[1]

[Total 2 marks]

2   Every element can be represented by a chemical symbol. These are all shown in the periodic table.

Remember — you can use a periodic table to help you answer any of these questions. There's one at the back of this book (you'll get one in your exams too).

a) State the chemical symbol for sodium.

....

[1]

b) The diagram below shows the position of the element with the symbol Bi in the periodic table.

Line A

Bi

i) Which element does the chemical symbol Bi represent?

....

[1]

ii) Element **X** is found in the same group of the periodic table as Bi.
What does that tell you about the properties of the two elements?

....

[1]

iii) Element **X** does not conduct electricity. Predict whether element **X** will be found to the left or the right of line **A** in the diagram above. Explain your answer.

....

....

[1]

[Total 4 marks]

Score:

6

Atomic Mass and Isotopes

1. Aluminium appears in the periodic table as shown below.

27
Al
13

a) State how many protons there are in an atom of aluminium.

....
[1]

b) i) What information does the mass number of an atom give you?

....
[1]

ii) How many neutrons are there in an atom of aluminium?

....
[1]

[Total 3 marks]

2. Two of the most common isotopes of chlorine are chlorine-35 (^{35}Cl) and chlorine-37 (^{37}Cl).

a) What is meant by the term *isotope*?

....
....
[1]

b) Complete the following table to show the mass number and the numbers of protons and neutrons in an atom of each of these chlorine isotopes.

Isotope	Mass number	Number of protons	Number of neutrons
^{35}Cl	35	17	
^{37}Cl			

[2]

c) The *relative atomic mass* of chlorine is 35.5.
What is meant by the term *relative atomic mass* (A$_r$)?

....
....
[2]

[Total 5 marks]

Score: [ ]

8

Electron Shells

1 Beryllium, magnesium and calcium are all Group 2 elements.

a) Give the electronic structure of beryllium. ....

[1]

b) The diagram on the left represents the electronic structure of a magnesium atom.
Complete the diagram on the right to show the electronic structure of a calcium atom.

Mg

Ca

[2]

c) State how many extra electrons an atom of magnesium would need to fill up its outer shell.

....

[1]

[Total 4 marks]

2 The electronic structure of an atom shows how many electrons there are in each energy level.

a) An atom has the electronic structure 2, 8, 8, 1. Identify which element the atom is.

....

[1]

b) The diagram on the right incorrectly shows the electronic structure of neon.

i) State what is wrong with the first energy level in the diagram.
Give a reason for your answer.

....

....

[2]

Ne

ii) State what is wrong with the second energy level in the diagram.
Give a reason for your answer.

....

....

[2]

[Total 5 marks]

Score:

9

Section One — Fundamental Ideas in Chemistry

Compounds

1 Bromine is a Group 7 element. It is a non-metal with seven electrons in its outer shell.

a) Hydrogen is a non-metal with one electron in its outer shell.
Describe how a hydrogen atom and a bromine atom react together to form a covalent compound.

....

....

....

[2]

b) Lithium is a metal with one electron in its outer shell.
Describe how a lithium atom and a bromine atom react together to form an ionic compound.

....

....

....

....

[3]

[Total 5 marks]

2 The diagram below shows how sodium and chlorine react to form a bond.

a) Explain why sodium reacts with chlorine to form a compound in this way.

....

....

[1]

b) Give the chemical formula of the compound produced.

....

[1]

[Total 2 marks]

Score:

7

Section One — Fundamental Ideas in Chemistry

Relative Formula Mass

1 A student is asked to calculate the relative formula mass (M_r) of the compound calcium sulfate, $CaSO_4$.

Use the periodic table to help you answer the questions on this page. There's one at the back of the book. You'll get one in the exam too.

a) What is the *relative formula mass* of a compound?

....

[1]

b) Find the relative formula mass (M_r) of calcium sulfate, $CaSO_4$.

....

[1]

[Total 2 marks]

2 A student is given the chemical formula and relative formula mass (M_r) of an unknown compound which contains an unknown element, **Z**, as shown below.

Chemical formula = Z_2CO_3 $M_r = 106$

a) Identify element **Z**.

....

....

....

....

[3]

b) State the mass in grams of one mole of the compound. .... g

[1]

c) The student has a 2.7 g sample of the compound. Calculate the number of moles in the sample.

Number of moles = ...

[1]

[Total 5 marks]

3 A teacher has a 140 g sample of potassium hydroxide (KOH). He needs a sample equivalent to 4 moles.

Calculate how much more KOH in grams the teacher needs in order to have a 4 mole sample.

Extra KOH needed = ... g

[Total 3 marks]

Score: [ ] **10**

Section One — Fundamental Ideas in Chemistry

Percentage by Mass and Empirical Formula

1   A student was asked to work out the percentage mass of different elements in compounds.

    a)   Calculate the percentage mass of nitrogen in ammonium nitrate, NH_4NO_3.

Nitrogen: .... %

[3]

    b)   Calculate the percentage mass of oxygen in iron(III) oxide, Fe_2O_3.

Oxygen: .... %

[3]

[Total 7 marks]

2   A compound contains 10.1% aluminium and 89.9% bromine by mass.

    Find the empirical formula of the compound.

Empirical formula = ...

[Total 3 marks]

3   1.48 g of a calcium compound contains 0.8 g of calcium and 0.64 g of oxygen.

    The rest of the compound is hydrogen. Find the empirical formula of the compound.

Empirical formula = ...

[Total 4 marks]

Score: ☐

13

Section One — Fundamental Ideas in Chemistry

Balancing Equations

1 Methane (CH_4) burns in oxygen (O_2) to make carbon dioxide (CO_2) and water (H_2O).

a) State the names of the reactants and products in this reaction.

Reactants: ....

Products: ....

[2]

b) Write a word equation for this reaction.

....

[1]

c) Write a balanced chemical equation for this reaction.

....

[2]

[Total 5 marks]

2 Sulfur (S) burns in oxygen (O_2) to make sulfur dioxide (SO_2).

$$S + O_2 \rightarrow SO_2$$

2 g of sulfur is required to produce 4 g of sulfur dioxide in the reaction shown above. State the mass of oxygen that will react with 2 g of sulfur. Explain your answer.

....

....

....

[Total 2 marks]

3 Acids can react with a variety of different metals and their oxides.

a) Balance the following chemical equations.

i) ....HCl + ...CuO → ....$CuCl_2$ + ...H_2O

[1]

Not every compound will need a number in front of it to balance these equations.

ii) ....HNO_3 + ...MgO → ....$Mg(NO_3)_2$ + ...H_2O

[1]

b) Write a balanced chemical equation for the reaction of hydrochloric acid (HCl) with aluminium (Al) that produces aluminium chloride ($AlCl_3$) and hydrogen (H_2).

....

[2]

[Total 4 marks]

4 A more reactive halogen can displace a less reactive halogen from a solution of its salt.

Write a balanced chemical equation for the reaction of chlorine (Cl_2) with potassium bromide (KBr). The products of this reaction are bromine (Br_2) and potassium chloride (KCl).

....

[Total 2 marks]

5 Reactions involving the removal of oxygen from a compound are called reduction reactions.

a) Balance the following equation, which shows the reduction of iron(III) oxide.

$$........CO + ...Fe_2O_3 \rightarrow ....CO_2 + ...Fe$$

[1]

b) Balance the following equation, which shows the reduction of copper(II) oxide.

$$........CuO + ...C \rightarrow ....Cu + ...CO_2$$

[1]

[Total 2 marks]

6 Sodium (Na) is a reactive alkali metal, which is found in Group 1 of the periodic table.

a) Sodium metal reacts with chlorine gas to form sodium chloride (NaCl).

Write a balanced chemical equation for this reaction.

....

[2]

b) When a solution of calcium hydroxide ($Ca(OH)_2$) is mixed with solid sodium carbonate (Na_2CO_3), sodium hydroxide solution (NaOH) and a precipitate of calcium carbonate ($CaCO_3$) are produced.

The equation for the reaction is:

$$Ca(OH)_2(.........) + Na_2CO_3(.........) \rightarrow 2NaOH(.........) + CaCO_3(.........)$$

Complete the equation by adding state symbols.

[1]

c) Solid sodium metal reacts with water to form a solution of sodium hydroxide (NaOH). Hydrogen gas (H_2) is also given off.

Write a balanced chemical equation for this reaction, including state symbols.

....

[3]

[Total 6 marks]

Exam Practice Tip

It's important that you get to grips with balancing equations because it often features in exam papers. Remember to double-check your equation after you've balanced it and you'll be on to a winner. You need to make sure you know your state symbols too, because they come hand in hand with chemical equations.

Score

[ ]
___
21

 [ ]   [ ]   [ ]

Section One — Fundamental Ideas in Chemistry

Calculating Masses in Reactions

1   A student is investigating the combustion of metals.

a) The student burns 10 g of magnesium in air to produce magnesium oxide. Use the equation
below to calculate the maximum mass of magnesium oxide that could be produced in the reaction.

$$2Mg \ + \ O_2 \ \rightarrow \ 2MgO$$

Mass of magnesium oxide = ... g

[3]

b) Using the chemical equation below, calculate the mass of sodium
that the student would need to burn in order to produce 2 g of sodium oxide.

$$4Na \ + \ O_2 \ \rightarrow \ 2Na_2O$$

Mass of sodium = ... g

[3]

[Total 6 marks]

2   Aluminium and iron(III) oxide (Fe_2O_3) react together to produce aluminium oxide (Al_2O_3)
and iron. The equation for the reaction is shown below.

$$Fe_2O_3 \ + \ 2Al \ \rightarrow \ Al_2O_3 \ + \ 2Fe$$

a) What is the maximum mass of iron that can be produced from 20 g of iron oxide?

Mass of iron = ... g

[3]

b) What is the maximum mass of aluminium that will react with 32 kg of iron(III) oxide?

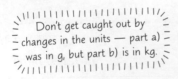
Don't get caught out by
changes in the units — part a)
was in g, but part b) is in kg.

Mass of aluminium = ... kg

[3]

[Total 6 marks]

3 Iron oxide is reduced to iron inside a blast furnace using carbon.
There are three stages involved. The equations for these three stages are shown below.

Stage 1: $C (s) + O_2 (g) \rightarrow CO_2 (g)$

Stage 2: $CO_2 (g) + C (s) \rightarrow 2CO (g)$

Stage 3: $3CO (g) + Fe_2O_3 (s) \rightarrow 3CO_2 (g) + 2Fe (l)$

If 10 g of carbon is used in stage 2, and all the carbon monoxide produced gets used in stage 3, what mass of CO_2 is produced in stage 3?

Mass of CO made from 10 g of C
at Stage 2:

C	2CO
12	2 × (12 + 16) = 56
12 ÷ .... = ... g	56 ÷ .... = ... g
.... × 10 = ... g	.... × 10 = ... g

Mass of CO_2 made from .... g of CO
at Stage 3:

3CO	3CO₂
3 × (12 + 16) = 84	3 × [12 + (16 × 2)] = 132
84 ÷ .... = ... g	132 ÷ .... = ... g
.... × .... = ... g	.... × .... = ... g

Mass of CO_2 = ... g

[Total 6 marks]

4 Sodium sulfate (Na_2SO_4) is made by reacting sodium hydroxide (NaOH) with sulfuric acid (H_2SO_4). Water is another product of this reaction. The equation for the reaction is shown below.

$$2NaOH + H_2SO_4 \rightarrow Na_2SO_4 + 2H_2O$$

a) What mass of sodium hydroxide is needed to make 75 g of sodium sulfate?

Mass of sodium hydroxide = ... g

[3]

b) What is the maximum mass of water that can be formed when 50 g of sulfuric acid reacts with sodium hydroxide?

Mass of water = ... g

[3]

[Total 6 marks]

Exam Practice Tip

It's really important that you know how to calculate masses in reactions. It's a big examiner favourite. The trickiest questions are ones where there's more than one mole of a reactant or product — watch out for them. And double-check your working when you're done, to make sure your answer is right and the units are correct.

Score

24

Section One — Fundamental Ideas in Chemistry

Percentage Yield and Reversible Reactions

1 A teacher wanted to produce some silver chloride (AgCl). He added a carefully measured mass of silver nitrate to an excess of dilute hydrochloric acid. 1.2 g of silver chloride was produced.

a) What is meant by the *yield* of a chemical reaction?

....
[1]

b) The teacher calculated that he should have got 2.7 g of silver chloride from the reaction. What was the percentage yield?

Percentage yield = ... %
[2]

[Total 3 marks]

2 The reaction between carbon monoxide (CO) and hydrogen (H_2) is reversible.

$$CO + 2H_2 \rightleftharpoons CH_3OH$$

a) What is a *reversible reaction*?

....
[1]

b) Explain why the yield of methanol (CH_3OH) from this reaction is never 100%.

....

....
[1]

[Total 2 marks]

3 Some students mixed together barium chloride ($BaCl_2$) and sodium sulfate (Na_2SO_4) solutions in a beaker. Insoluble barium sulfate was formed. They filtered the solution to obtain the solid substance, then transferred the solid to a clean piece of filter paper to dry.

a) The students calculated that they should obtain 15 g of barium sulfate. Their actual yield was 6 g. Calculate the percentage yield of barium sulfate for this reaction.

Percentage yield = ... %
[1]

b) Suggest **one** reason why their actual yield was lower than their predicted yield.

....

....
[1]

[Total 2 marks]

Score: [ ]

7

Compounds, Ions and Ionic Bonding

1 Many substances, such as copper, are elements. Others, such as table salt, are compounds.

Briefly describe what a *compound* is.

....

....

[Total 2 marks]

2 The electronic structures of argon, fluorine and magnesium are shown below.

argon   fluorine   magnesium

a) Give the name of the element above that forms ions by losing electrons. ....

[1]

b) Give the name of the element above that forms negative ions. ....

[1]

c) Give the name of the element above that will not form an ion. Explain your answer.

....

....

....

[3]

[Total 5 marks]

3 An atom of a Group 6 element forms a stable ion by gaining electrons.

a) Explain how you can tell that the element must be a non-metal.

....

[1]

b) State the charge that the ion will have. ....

[1]

[Total 2 marks]

4 Describe what happens when a chlorine atom reacts with a lithium atom.

....

....

....

[Total 3 marks]

5   Sodium chloride has a giant ionic structure.

a) What is the name given to the arrangement of ions in a giant ionic structure?

....

[1]

b) Explain how the ions in a giant ionic structure are held together.

....

....

....

[3]

c) Draw a diagram showing the electron arrangements of the ions in sodium chloride.
Include the charge on each ion in your diagram.

Sodium chloride is formed when a sodium atom donates one electron to a chlorine atom.

[3]

[Total 7 marks]

6   A student did an experiment to see if an ionic compound conducted electricity when solid and when dissolved in water. The results are shown in the table below.

	Conducts electricity?
When solid	No
When dissolved in water	Yes

a) Explain the student's results in terms of ions.

....

....

....

[3]

b) Ionic compounds also conduct electricity when they are melted, but the student couldn't get the compound hot enough to melt it. Explain why ionic compounds have high melting points.

....

....

[2]

[Total 5 marks]

Score: ☐

24

Section Two — Bonding and Structure

Covalent Bonding

1 Atoms in molecules are held together by covalent bonds.

a) State what a *covalent bond* is and explain why atoms form covalent bonds.

....

....

....
[2]

b) Three students were asked to draw diagrams to represent the bonding in a molecule of hydrogen (H₂). The diagrams they drew are shown below.

H=H   [H | H]⁺⁻   H—H

Diagram A   Diagram B   Diagram C

Give the letter of the diagram that correctly represents the bonding in H₂. ....
[1]
[Total 3 marks]

2 Dot and cross diagrams can be used to show the position of electrons in covalent molecules.

a) Complete the dot and cross diagrams for the molecules below. Only show the outer electrons.

i) hydrogen chloride, HCl   ii) oxygen, O₂   iii) ammonia, NH₃

[1]   *[1]*   *[1]*

b) Describe the bonding in a molecule of hydrogen chloride.

....

....

....
[2]
[Total 5 marks]

3 Methane (CH_4), ammonia (NH_3) and oxygen (O_2) are all covalently bonded molecules.

a) i) Draw a dot and cross diagram to show the bonding in a molecule of methane.

[2]

ii) How many covalent bonds are present in one molecule of methane? ....
[1]

b) How many covalent bonds are present in one molecule of ammonia? ....
[1]

c) Choose words from the box to complete the sentences below.
Each word may be used once, more than once, or not at all.

metal	methane	cation	ammonia	oxygen	non-metal

i) Some molecules, such as .... contain a double covalent bond.

[1]

ii) A compound formed when a non-metal reacts with a .... will be
made up of covalently bonded molecules.

[1]
[Total 6 marks]

4 The atoms in hydrogen sulfide (H_2S) are bonded in a similar way to the atoms in water (H_2O).

Draw a dot and cross diagram of hydrogen sulfide. Only show the outer electrons of each atom.

Sulfur is in the same group of the periodic table as oxygen (Group 6), so it has the same
number of outer electrons.

[Total 2 marks]

Exam Practice Tip
The crucial thing to remember here is that covalent bonds form so that each atom has a full outer shell of
electrons. Also, make sure you know how many electrons each shell can hold. If you can get those two things
fixed in your head you should be able to work out the bonding of any molecules you're given in the exam.

Score

16

Section Two — Bonding and Structure

Covalent Substances: Two Kinds

1 The table below shows the properties of four substances.

Substance	Melting point (°C)	Conducts electricity when a liquid
A	−102	no
B	1085	yes
C	993	yes
D	1610	no

a) State and explain which substance could be silicon dioxide.

Substance ....

Explanation ....

....

[3]

b) State and explain which substance could be chlorine (Cl$_2$).

Substance ....

Explanation ....

....

[3]

[Total 6 marks]

2 Methane is a simple molecular substance. It is a gas at room temperature.

a) Explain why simple molecular substances like methane have low boiling points.

....

....

....

[2]

b) Explain why methane does not conduct electricity.

....

....

[1]

[Total 3 marks]

3   Graphite, diamond and silicon dioxide are all substances with giant covalent structures.

a) *In this question you will be assessed on the quality of your English,*
the organisation of your ideas and your use of appropriate specialist vocabulary.

The table below contains information about some of the properties of diamond and graphite.

	Hardness	**Melting point**	**Conducts electricity?**
Diamond	Hard	High	No
Graphite	Soft	High	Yes

Explain these properties of diamond and graphite in terms of their structure and bonding.

....

....

....

....

....

....

....

....

....

....

....

....

....

[6]

b) The diagrams below show the arrangement of the atoms in three different compounds.

A   B   C

Which diagram, **A**, **B** or **C**, shows the arrangement of atoms in silicon dioxide? ....

[1]

[Total 7 marks]

Score: ☐

16

😐 ☐   🙂 ☐   😊 ☐

Fullerenes and Nanoscience

1 Fullerenes are a type of nanoparticle.

a) Complete the sentences below.

Fullerenes are made up of .... atoms.

These atoms are arranged into rings in the shape of a ....

[2]

b) i) Fullerenes can be joined together to form nanotubes. Give **one** use of nanotubes.

....

[1]

ii) Give **two** other uses of fullerenes.

1 ....

2 ....

[2]

[Total 5 marks]

2 Nanoparticles often have different properties to the bulk material.

a) Zinc oxide is good at absorbing UV rays, so it is used in many sun creams.
Zinc oxide powder is white, but zinc oxide nanoparticles are transparent.

i) Suggest why a company that makes sun cream may decide to start using
zinc oxide nanoparticles in their product instead of zinc oxide powder.

....

....

[2]

ii) Draw a ring around the most likely size of zinc oxide nanoparticles, from the options below.

0.1 nm   **30 nm**   **1000 nm**   **200 nm**

[1]

b) Many industrial processes rely on transition metal catalysts. In the future new, more effective
versions of these catalysts may be developed, using nanoparticles of the same metals.

Suggest why nanoparticles might make more effective catalysts than the same materials in bulk.

....

....

[1]

[Total 4 marks]

Score:

9

Section Two — Bonding and Structure

Section Three — Air and Water

Air

1   This pie chart shows the composition of air.

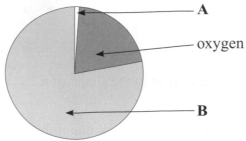

a)   Name the gas represented by segment **B**.

....
                                                                                                                                                        [1]

b)   What percentage of air is composed of oxygen?

....
                                                                                                                                                        [1]

c)   Segment **A** represents all of the gases that are only present in air in small amounts.
     Name **two** of the gases that are represented by segment **A**.

1 ....

2 ....
                                                                                                                                                        [2]

[Total 4 marks]

2   A chemical plant obtains oxygen gas by separating out the gases that make up the air.

a)   Name the process that is used to separate out the gases in air.

....
                                                                                                                                                        [1]

b)   In the plant, the air is cooled to turn it into a liquid and then passed into a fractionating column.
     What property of the substances in the air causes them to separate out in the fractionating column?

....
                                                                                                                                                        [1]

c)   Describe a laboratory test you could use to identify oxygen gas.

....

....
                                                                                                                                                        [2]

[Total 4 marks]

Score: [   ]

8

Oxygen and Burning

1 Some elements burn in air to produce oxides, which may be alkaline or acidic.

a) Complete this table about the reaction of certain elements with oxygen.

Element	Flame colour when burnt	Oxide formed	Acid-base character of oxide
sodium	....	Na_2O	....
magnesium	....	....	Alkaline
carbon	Orange/yellow	CO_2	....
sulfur	Blue	....	Acidic

[6]

b) Aluminium oxide is *amphoteric*. State what is meant by *amphoteric*.

....

[1]

[Total 7 marks]

2 When iron burns in air, iron oxide is formed.

a) Write the word equation for this reaction.

....

[1]

b) State whether iron is oxidised or reduced during this reaction and explain your answer.

....

....

....

[2]

[Total 3 marks]

3 Potassium is a metal which burns in air to form potassium oxide.

When potassium oxide dissolves in water, will the solution formed be acidic, neutral or alkaline? Explain your answer.

....

....

....

[Total 2 marks]

Score: ☐

12

 ☐   ☐   ☐

Air Pollution

1 Fumes from faulty central heating boilers can contain carbon monoxide.

a) What can cause carbon monoxide to be produced when fuel is burnt in a boiler?

....

[1]

b) Explain why carbon monoxide is poisonous.

....

....

[2]

[Total 3 marks]

2 Oxides of nitrogen are formed when nitrogen and oxygen in the air react at high temperatures.

a) Give an example of where this reaction might take place.

....

[1]

b) The equation below shows nitrogen dioxide reacting with moisture in the atmosphere.
Complete the equation so that it is balanced.

$$.......... NO_2 + H_2O \rightarrow .... HNO_3 + NO$$

[2]

[Total 3 marks]

3 An air quality monitoring station recorded a large increase in the level of sulfur dioxide in the air shortly after a new coal-fired power station started operating nearby.

a) Explain why burning fossil fuels produces sulfur dioxide.

....

....

[2]

b) A conservation charity is worried about the effect that high sulfur dioxide levels in the air might have on the wildlife in a local lake. Suggest why they are concerned.

....

....

....

[3]

[Total 5 marks]

Score:

11

Water Quality

1 A student has a sample of a liquid. He wants to know whether it contains water.

a) Describe a chemical test that the student could use to find out whether the sample contains water. State what the result would be if the sample does contain water.

....

....

....

[2]

b) He then boils the liquid.
What will its boiling point be if the sample is pure water?

....

[1]

[Total 3 marks]

2 In some countries, a process called *desalination* is used to produce water for use in households.

a) i) What is meant by *desalination*?

....

[1]

ii) Suggest why this process may be important in countries with low rainfall.

....

....

[2]

b) Name another method that can be used to produce pure water.

....

[1]

[Total 4 marks]

3 Before water can be used in homes, it must be treated to ensure it is clean.

Describe how water from a reservoir would be treated at a treatment works to prepare it for use in homes. Explain what each stage of the process is for.

....

....

....

....

[Total 4 marks]

Section Three — Air and Water

4 Drinking water often has substances such as chlorine and fluoride added to it.

*In this question, you will be assessed on the quality of your English,
the organisation of your ideas and your use of appropriate specialist vocabulary.*

Evaluate the advantages and disadvantages of adding chlorine and fluoride to the supply of
drinking water. Remember to finish your answer with a conclusion.

....

....

....

....

....

....

....

....

....

....

[Total 6 marks]

5 Many people buy water filters to treat their tap water.

a) Give **two** reasons why someone might use a water filter containing carbon or silver.

1 ....

2 ....

[2]

b) Water filters that use ion exchange resins are used to reduce water hardness.
Explain how a water filter based on an ion exchange resin works.

....

....

....

....

[3]

[Total 5 marks]

Exam Practice Tip

If you get a question asking you to evaluate something, there are a few things you need to do in order to give yourself a chance at getting full marks. Firstly, you have to give points for both sides of the argument (the for and the against), and secondly, you need to finish up with the conclusion you've drawn from your argument.

Score

22

Rust

1 In an experiment to investigate rusting, three iron nails were placed in separate test tubes.

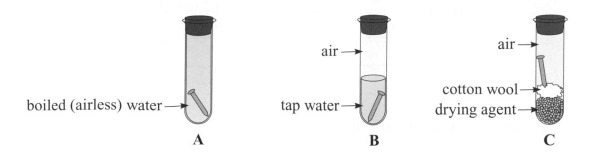

a) Complete the sentences using words from the box below.

corrosion	water	reduction	acid	dissolving	zinc	air

Rusting is the .... of iron.

For iron to rust both .... and .... must be present.

[3]

b) In which tube, **A**, **B** or **C**, will the nail rust?

....

[1]

c) Experiment **B** was repeated, but using a nail that had first been coated in paint.
Suggest what you would observe in this experiment and explain your answer.

....

....

[2]

d) One method of protecting iron nails from rusting is by coating them with zinc.
This method is called sacrificial protection.

Explain how this method prevents iron from rusting even if the coating is scratched
and the iron underneath is revealed.

....

....

[2]

e) Suggest a method that could be used to prevent moving parts in a machine from rusting.

....

[1]

[Total 9 marks]

Score:

9

Section Three — Air and Water

More About The Periodic Table

1   The table below shows the electronic structures of fluorine and chlorine.

Element	Electronic structure
F	2, 7
Cl	2, 8, 7

Use this information to explain why fluorine is more reactive than chlorine.

....

....

....

....

....

[Total 4 marks]

2   The periodic table contains patterns in the properties of the elements.

a) In what order are the elements in the periodic table arranged?

....

[1]

b) Tick **one** box to show which pair of elements have similar properties.

☐ C and O   ☐ He and F   ☐ Si and Mg   ☐ Cl and Br

[1]

c) The reactions of sodium are very similar to the reactions of potassium. Explain why this is.

....

....

[1]

d) Explain why the noble gases (Group 0 elements) are unreactive.

....

....

[2]

[Total 4 marks]

Exam Practice Tip

Questions about how the reactivity of the elements varies as you go down a group are old examiner favourites, so it's worth making sure you've got your head around this topic. Remember, it's not just about learning the patterns in reactivity — you need to be able to explain <u>why</u> the reactivity changes as you move down a group.

Score

☐

__9__

 ☐   ☐

Group 1 — The Alkali Metals

1   A student put a piece of lithium into a beaker of water and observed that a reaction occurred.

a) The piece of lithium floated on top of the water. Explain why.

....

[1]

b) After the reaction had finished, the student tested the pH of the solution with universal indicator.
   Would the solution be acidic, alkaline or neutral? Draw a ring around the correct word.

          acidic   **alkaline**   **neutral**

[1]

c) Complete and balance the symbol equation for this reaction.

   $$\text{........ Li (s)} + \text{........ H}_2\text{O (l)} \rightarrow 2\,\text{LiOH (aq)} + \text{................. (g)}$$

[2]

[Total 4 marks]

2   Explain why the alkali metals become more reactive as their atomic number increases.

....

....

....

[Total 4 marks]

3   Sodium metal will react with iodine gas to form sodium iodide.

$$2\text{Na} + \text{I}_2 \rightarrow 2\text{NaI}$$

a) Describe the appearance of sodium iodide and state the charge on the sodium ion
   in this compound.

....

....

[2]

b) Describe what you would observe if some sodium iodide was placed in a beaker of water.

....

....

....

[2]

[Total 4 marks]

Score:

12

Group 7 — The Halogens

1 When iron reacts with bromine a solid red-brown compound forms on the sides of the test tube.

a) Give the name of the solid red-brown compound and state what type of bonding is present within this compound.

....

....

[2]

b) State the charge on the bromide ion in this compound.

....

[1]

[Total 3 marks]

2 The table below shows the physical properties of two Group 7 elements.

Element	Melting Point (°C)	Boiling Point (°C)
chlorine	−101.5	−34.0
bromine	−7.3	58.8

a) Name a halogen with a lower boiling point than chlorine.
Justify your answer in terms of the trends in Group 7.

....

....

[3]

b) Name a halogen that is less reactive than bromine. ....

[1]

[Total 4 marks]

3 When bromine water is added to potassium iodide solution a reaction will take place.

a) Write a balanced symbol equation for this reaction.

....

[2]

b) Explain why this reaction happens.

....

....

[2]

[Total 4 marks]

Score:

11

Transition Elements

1 The transition elements and the Group 1 elements are metals. A teacher carried out tests on samples of metals from both groups to demonstrate their properties to some students.

For each of the following tests, say which metal is most likely to be the transition element and explain your answer.

a) Metal **A** could be cut with a knife, but metal **B** could not.

....

....

[1]

b) Metal **C** melted when heated with a Bunsen burner, but metal **D** did not.

....

....

[1]

c) Metals **E** and **F** were placed in water. Metal **E** sank and no immediate reaction was observed. Metal **F** floated and fizzed vigorously.

....

....

[2]

[Total 4 marks]

2 Iron is a transition element. Iron and its compounds have many uses.

a) Iron is added to the reaction vessel during the Haber process, in which nitrogen reacts with hydrogen to produce ammonia. Suggest what role the iron is playing in this reaction.

....

[1]

b) Iron oxide can exist in different forms, for example FeO and Fe_2O_3.

i) Suggest why iron oxide can exist as either FeO or Fe_2O_3.

....

....

[1]

ii) Some artists' paints contain Fe_2O_3.
Suggest what property of this compound makes it suitable for use in paints.

....

[1]

[Total 3 marks]

Score:

7

Section Four — The Periodic Table and Metals

Reactions of Metals and the Reactivity Series

1   Four different metals are reacted with dilute sulfuric acid.
     The diagram below shows the reactions after 30 seconds.

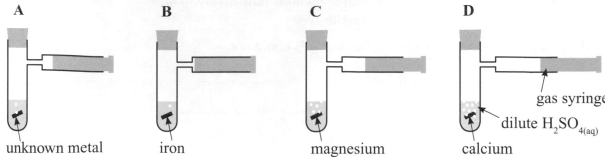

a) i) Name the **two** products of reaction **C**.

    1 ....

    2 ....
                                                                                                              [2]

    ii) Describe a simple test that you could perform to confirm the identity of the gas
        collected from reaction **C**.

        ....

        ....
                                                                                                              [2]

b) Write the letters **A-D** in the spaces below to arrange the reactions in order of reactivity.

    Least vigorous ────────────────────────────────▶ **Most vigorous**

    Reaction .... → Reaction .... → Reaction .... → Reaction ....
                                                                                                              [2]

c) Use your knowledge of the reactivity series and your answer to part b) to suggest which metal is
    the unknown metal in the diagram above.

    Write out part of the reactivity series for metals...

    More reactive: | Calcium
                   | Magnesium
                   | Zinc
                   | Iron
    Less reactive: ▼ Copper

    Unknown metal: ....
                                                                                                              [1]

d) Complete the word equation for the reaction of sodium with water.

    sodium + water → .... + ...
                                                                                                              [2]
                                                                                                    [Total 9 marks]

2 A student placed pieces of copper, zinc and another unknown metal in zinc sulfate solution and in copper sulfate solution and left them for an hour. The student's results are shown below.

	zinc	copper	unknown metal
reaction with zinc sulfate	no reaction	no reaction	no reaction
reaction with copper sulfate	reaction	no reaction	reaction

a) Suggest the name of the unknown metal.

....

[1]

> Writing out the reactivity series on a spare bit of paper will help with this question.

b) Explain how you can tell that the unknown metal is more reactive than copper.

....

....

[2]

c) When zinc is added to copper sulfate the copper ions are reduced by the zinc.
Complete the ionic equation for this reaction.

$$Cu^{2+} (aq) \; + \; Zn (s) \rightarrow \; .... (aq) \; + \; .... (s)$$

[1]

[Total 3 marks]

3 Hydrogen and carbon are often included in the reactivity series.

More reactive: | Magnesium
           | Carbon
           | Zinc
           | Iron
Less reactive: ↓ | Hydrogen

a) Magnesium is above carbon in the reactivity series.
State what you would observe if a sample of magnesium oxide was heated with carbon.

....

[1]

b) Iron reacts with hydrochloric acid to produce iron chloride and a gas.

i) Name the gas that is produced.

....

[1]

ii) Copper will not react with hydrochloric acid to form copper chloride.
Explain why iron does react with the acid but copper does not.

....

....

[2]

[Total 4 marks]

Score: ☐

16

 ☐ ☐ ☐

Getting Metals From Rocks

1   Some metals can be extracted by heating their ore with carbon.

a) Explain why iron can be extracted from iron oxide by heating with carbon.

....

....

....

[2]

b) What is the name given to this type of reaction?

....

[1]

c) Complete the word equation for the reaction of iron(III) oxide with carbon.

iron(III) oxide + ... → .... + ...

[1]

d) Explain why gold does not need to be extracted using carbon.

....

....

[2]

[Total 6 marks]

2   Not all metals can be extracted using carbon.
     Some need to be extracted using a different method.

a) i) Give the name of the process used to extract these metals.

....

[1]

ii) Explain the main disadvantage of using this process to extract metals.

....

....

[2]

b) Explain why not all metals can be extracted using carbon.

....

....

[1]

[Total 4 marks]

3   Malachite is a naturally occurring copper ore. Copper can be
extracted from malachite by heating the ore in a furnace with carbon.

a) Give the name for the process of extracting a metal by heating its ore in a furnace.

....

[1]

b) Suggest why the copper produced by this method cannot be used to make electrical wires.

....

....

[2]

[Total 3 marks]

4   Copper can be purified by electrolysis.

a) Identify the components of the electrolysis cell used for this process by writing the correct letter on each of the lines below.

copper sulfate solution   ....

copper ions   ....

pure copper electrode   ....

impure copper electrode   ....

[2]

b) Write down the half equations for the reactions at:

i) the anode. ....

[1]

ii) the cathode. ....

[1]

c) The experiment was repeated with two new electrodes, which were weighed before and after the experiment. The table below shows the results.

	Electrode X (mass in g)	Electrode Y (mass in g)
Before	122.6	25.1
After	34.0	113.9

Which electrode was used as the anode? Explain your answer.

....

....

[2]

[Total 6 marks]

Section Four — The Periodic Table and Metals

5 The equation below shows the reaction that occurs when scrap iron is heated with copper sulfate solution.

$$iron + copper\ sulfate \rightarrow copper + iron\ sulfate$$

a) Why does this displacement reaction take place?

....

....

[1]

b) Write an ionic equation for the reaction between the iron atoms and the copper ions.

....

[2]

c) Which substance is oxidised during this reaction?

....

[1]

[Total 4 marks]

6 Bioleaching is one method that can be used to extract copper from copper ores. It has a lower environmental impact than more traditional methods.

a) Briefly describe how bioleaching can be used to extract copper from copper sulfide.

....

....

....

[2]

b) Name another copper extraction method with a low environmental impact.

....

[1]

c) Apart from their low environmental impact, explain why these alternative extraction methods are important.

....

....

....

....

[2]

[Total 5 marks]

Exam Practice Tip

There are two main metal extraction processes: reduction with carbon and electrolysis. But you also need to know about the fiddly little alternative methods for extracting copper. Knowing the reactivity series will really help you answer questions about metal extraction too, so try looking over it again if you're struggling.

Score

28

Section Four — The Periodic Table and Metals

Impacts of Extracting Metals

1 Metal-rich ores are obtained by mining. The metals can then be extracted by reduction with carbon or by electrolysis.

a) *In this question you will be assessed on the quality of your English, the organisation of your ideas and your use of appropriate specialist vocabulary.*

Evaluate the advantages and disadvantages of mining metal ores.
Remember to finish your answer with a conclusion.

....

....

....

....

....

....

....

....

....

....

....

....

[6]

b) An alternative method of obtaining metals is to recycle old metal items.
Give **three** advantages of recycling metals.

1. ....

....

2. ....

....

3. ....

....

[3]

[Total 9 marks]

Exam Practice Tip

DON'T PANIC — that's the most important thing to remember when you see one of these big questions. You'll be marked on how well you structure your answer, so take some time to think about the points you're going to make. And if you're asked for a conclusion, don't forget to write one, or you're throwing away marks.

Score

9

Metals

1   All metals have a similar structure.

a) Complete these sentences about the structure of metals.

Metals consist of giant structures of atoms arranged in a regular ....

The electrons in the outer shells of the metal atoms are ....

[2]

b) Draw a labelled diagram showing the structure of a typical metal.

[2]

c) Use your diagram from part b) to explain:

i) how metals conduct electricity.

....

....

[2]

ii) why metals can be bent and shaped.

....

....

[2]

[Total 8 marks]

2   Most metals that are used to make everyday objects are found in the
central section of the periodic table, which is shaded in the diagram below.

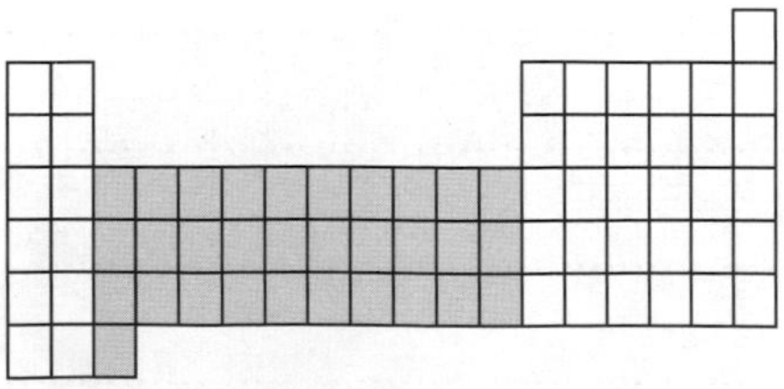

a) What name is given to this group of metals? ....

[1]

b) What properties of a metal from this group would make it suitable for making electrical wires?

....

....

[2]

[Total 3 marks]

3 Copper is a metal that is used in many everyday applications.

a) Suggest **two** properties of copper that make it suitable for making water tanks.

1. ....

2. ....

[2]

b) Cast iron has fewer uses than copper because it breaks easily. Complete the sentence.

Cast iron contains .... , which make the metal brittle.

[1]

[Total 3 marks]

4 A student carried out an experiment using the apparatus shown below. Four identically sized rods of different materials (**A**, **B**, **C** and **D**) were heated at one end. Temperature sensors were connected to the unheated ends. The student recorded the temperature every ten seconds.

temperature
sensor

HEAT

rod of
test material

The student's results are shown in the graph below.

Temperature of unheated
end of rod (°C)

A   B   C   D

Time (s)

a) Two rods were made of metals. Give the letters of the **two** rods that are most likely to be metal.

....

[1]

b) State which material, **A**, **B**, **C** or **D**, was the best conductor of heat. Explain your answer.

....

....

[2]

[Total 3 marks]

Score:

17

Alloys

1   Copper is a pure metal. Brass is an alloy.

a) What is an *alloy*?

....

[1]

b) Screws can be made from brass. Copper cannot be used to make screws because it is too soft. Explain why copper is softer than brass.

....

....

....

....

....

[4]

[Total 5 marks]

2   Steel is an alloy that is made by adding carbon to iron.

Complete the sentences.

a) Steels with .... amounts of added carbon are easily shaped.

[1 mark]

b) .... carbon steels are hard.

[1 mark]

c) .... steels are resistant to corrosion.

[1 mark]

[Total 3 marks]

3   Nitinol is a shape memory alloy.

a) Explain what is meant by the term *shape memory alloy*.

....

....

[2]

b) Explain why nitinol is used to make dental braces.

....

....

[2]

[Total 4 marks]

Score:

12

Section Four — The Periodic Table and Metals

Acids and Bases

1 The pH scale shows how acidic or alkaline substances are.

a) What range of values can pH take? ....

[1]

b) What term is used to describe a substance with a pH of 7? ....

[1]

c) Caustic soda has a pH of around 13. Which of these best describes caustic soda? Tick **one** box.

☐ strong   ☐ strong   ☐ weak   ☐ weak
  acid   alkali   acid   alkali

[1]

d) Universal indicator was added to a colourless solution with a pH of 3.
   What colour did the indicator turn? Draw a ring around the correct answer.

orange   **green**   **blue**   **purple**

[1]

[Total 4 marks]

2 A student added a few drops of universal indicator to a test tube containing some acid.
The solution became red. The student then gradually added an alkali to the test tube.

a) What type of ions in the acid caused the indicator to turn red? ....

[1]

b) i) What type of reaction took place between the acid and the alkali?

   ....

[1]

   ii) Write a chemical equation for the reaction that took place between the ions in the acid
       and the ions in the alkali. Include state symbols.

   ....

[2]

c) Apart from water, what type of chemical is formed in a reaction between an acid and a base?

....

[1]

d) The student stopped adding alkali when all of the acid had reacted.
   How could the student tell when all of the acid had reacted?

....

[1]

[Total 6 marks]

Score: ☐

10

Oxides, Hydroxides and Ammonia

1 Complete the word equations for the following reactions between acids and bases.

a) sulfuric acid + magnesium oxide → .... + ...

[1]

b) hydrochloric acid + aluminium hydroxide → .... + ...

[1]

[Total 2 marks]

2 Use words from the box to complete the sentences about ammonia.

| flavourings | insoluble | adhesives | fertilisers | acidic | alkaline |

Ammonia dissolves in water to form an .... solution.

Ammonia is used to make ammonium salts — these are widely used as ....

[Total 2 marks]

3 A student is investigating the reactions of acids.

a) The student reacts nitric acid with copper oxide. Write a word equation for the reaction.

....

[1]

b) The student adds zinc oxide (ZnO) to a test tube of hydrochloric acid (HCl).
Write a chemical equation for this reaction.

....

[2]

c) The student reacts sodium hydroxide with an acid to form sodium sulfate. Name the acid used.

....

[1]

[Total 4 marks]

4 The table below gives information on the solubility of five metal compounds.

a) Identify **one** compound from the table that is an alkali.

....

[1]

b) Identify **three** compounds from the table that are bases.

....

....

[1]

[Total 2 marks]

Compound	Soluble in water?
Iron hydroxide	no
Iron oxide	no
Lithium hydroxide	yes
Lithium chloride	yes
Silver chloride	no

Score:

10

Titrations

1   Nitric acid reacts with aqueous sodium
hydroxide to form a neutral solution.

A student wanted to find out how much
nitric acid is needed to neutralise a
25 cm³ sample of sodium hydroxide solution.
She decided to carry out a titration.

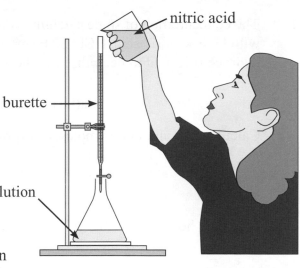

nitric acid

burette

sodium hydroxide solution
and indicator

a) Identify **one** potential hazard with the method shown
and suggest a way of minimising the risk to the student.

....

....
[2]

b) Universal indicator is not a suitable indicator to use for this experiment. Explain why.

....

....
[2]

c) As the student approached the end-point of the titration, she added the acid one drop at a time.
Suggest why she did this.

....

....
[1]

d) The volume of nitric acid required to neutralise the sodium hydroxide solution is the titre volume.

i) Describe how the student would take a titre volume reading from a burette.

....

....
[2]

ii) The student repeated the titration three times and used her results
to calculate a mean titre volume. Explain why she did this.

....

....
[1]

[Total 8 marks]

Score:

8

Titration Calculations

1 The concentration of some calcium hydroxide solution, $Ca(OH)_2$, was determined by titration with hydrochloric acid, HCl. 50.0 cm³ of limewater required 20.0 cm³ of 0.100 mol/dm³ hydrochloric acid to neutralise it. The equation for this reaction is shown below.

$$Ca(OH)_2 + 2HCl \rightarrow CaCl_2 + 2H_2O$$

a) Calculate the amount, in moles, of hydrochloric acid used in the reaction.

.... moles
[2]

b) Calculate how many moles of calcium hydroxide were used.

.... moles
[1]

c) Calculate the concentration of the limewater in mol/dm³.

Concentration = ... mol/dm³
[2]

d) Calculate the concentration of the limewater in g/dm³.

Concentration = ... g/dm³
[2]

[Total 7 marks]

2 In a titration, 10.0 cm³ of sulfuric acid was used to neutralise 30.0 cm³ of 0.100 mol/dm³ potassium hydroxide solution. The equation for the reaction is:
$$H_2SO_4 + 2KOH \rightarrow K_2SO_4 + 2H_2O$$

a) Calculate the concentration of the sulfuric acid in mol/dm³.

Moles of KOH = ... × (................ ÷ 1000) = ...

Reaction equation shows that .... moles of KOH reacts with 1 mole of H_2SO_4

So .... moles of KOH reacts with .... ÷ .... = ... moles of H_2SO_4

Concentration = ... ÷ (............... ÷ 1000)

= ....

Concentration = ... mol/dm³
[5]

b) Calculate the concentration of the sulfuric acid in g/dm³.

Concentration = ... g/dm³
[2]

[Total 7 marks]

Score: ☐
14

Making Salts

1 **A**, **B**, **C** and **D** are chemical equations for reactions in which salts are formed.

> **A** $CuO (s) + H_2SO_4 (aq) \rightarrow CuSO_4 (aq) + H_2O (l)$
>
> **B** $2NaOH (aq) + H_2SO_4 (aq) \rightarrow Na_2SO_4 (aq) + 2H_2O (l)$
>
> **C** $Pb(NO_3)_2 (aq) + H_2SO_4 (aq) \rightarrow PbSO_4 (s) + 2HNO_3 (aq)$
>
> **D** $Mg (s) + 2HCl (aq) \rightarrow MgCl_2 (aq) + H_2 (g)$

a) State which equation (**A** - **D**) shows the formation of a salt:

 i) in a reaction between an acid and an alkali. ....

 ii) by precipitation. ....

 iii)in a reaction between an acid and an insoluble base. ....

 iv)in a reaction between an acid and a metal. ....

[4]

b) A student proposes making the salt sodium chloride (NaCl) in the school laboratory by adding sodium metal to hydrochloric acid (HCl). Suggest why this is not an appropriate method to use.

 ....

 ....

[1]

[Total 5 marks]

2 Silver chloride is an insoluble salt which is formed as a precipitate when silver nitrate and sodium chloride solutions are mixed together.

a) Complete the word equation for the reaction.

 .... + ... $\rightarrow$ silver chloride + ...

[1]

b) Describe the steps that should be taken after mixing the solutions to produce a dry sample of silver chloride.

 ....

 ....

[2]

c) Precipitation reactions can be used to remove unwanted ions from solutions. An example of this is in the treatment of effluent. Give **one** other example.

 ....

[1]

[Total 4 marks]

3   Zinc nitrate is a soluble salt. It can be made by mixing an excess of zinc oxide powder with nitric acid, as shown in the diagram. Once the reaction is complete, the excess zinc oxide is separated from the zinc nitrate solution by filtration.

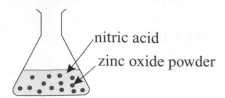

nitric acid

zinc oxide powder

a) Give **one** observation that would indicate that the reaction is complete.

....

....

[1]

b) Write a word equation for the reaction.

....

[1]

c) Describe how you could produce crystals of zinc nitrate from zinc nitrate solution.

....

....

[2]

d) Sodium nitrate is soluble. It can be made by reacting sodium hydroxide solution with nitric acid.

   i) Explain why the same method used for making zinc nitrate (as above) cannot be used.

....

....

....

[2]

   ii) Briefly describe how you would produce a pure solution of sodium nitrate from sodium hydroxide and nitric acid.

....

....

....

....

....

[3]

[Total 9 marks]

Exam Practice Tip

The two most important things to remember for this topic are (drumroll please): 1) To make any given salt, you need two substances that contain the right elements (e.g. to make lead chloride, you need lead ions and chloride ions). 2) The method you use depends on whether the salt *and* the base you're using to make it are soluble.

Score

[ ]

———

18

 [ ]   [ ]   🙂 [ ]

Metal Carbonates and Limestone

1 Areas of limestone can be found throughout the UK.
Limestone has uses in farming and in the building industry.

a) State the chemical name for the main compound in limestone.

....
[1]

b) How is limestone usually obtained?

....
[1]

c) When limestone is heated, it forms solid lime and one other product.

i) State the chemical name for lime.

....
[1]

ii) Name the other product that forms.

....
[1]

iii)Give the name for this type of reaction.

....
[1]

d) Limestone will react with acids. Complete the word equation below to show all the products formed in the reaction between limestone and an acid.

limestone + acid → salt + ... + ...
[2]

e) i) Explain why powdered limestone can be useful in farming.

....

....

....
[2]

ii) Give **two** other uses of limestone.

1. ....

2. ....
[2]

[Total 11 marks]

50

2 A student mixed a sample of magnesium carbonate with nitric acid. She collected the gas that was given off as the compounds reacted.

a) Name the salt that was formed by this reaction.

....

[1]

b) The student believes that the gas she has collected is carbon dioxide. Describe a simple laboratory test that the student could use to confirm this.

....

....

....

[2]

[Total 3 marks]

3 A student used a Bunsen burner to heat samples of copper carbonate and sodium carbonate.

a) i) Write a symbol equation for the reaction that occurs when copper carbonate is heated.

....

[2]

ii) Name the solid product of this reaction.

....

[1]

b) When the student heated the copper carbonate, a gas was given off. The student bubbled this gas through a solution of calcium hydroxide ($Ca(OH)_2$).

i) What is the name given to a solution of calcium hydroxide?

....

[1]

ii) The solution turned cloudy. Complete the symbol equation for the reaction that took place.

$$Ca(OH)_2 \ + \ .... \ \rightarrow \ .... \ + \ H_2O$$

[2]

c) Sodium is in Group 1 of the periodic table. When the student heated the sodium carbonate using a Bunsen burner, no gas was given off. Suggest why not.

....

....

....

[2]

[Total 8 marks]

Exam Practice Tip	Score
Don't get word and symbol equations mixed up — read the question carefully to make sure you give the right one. If you're asked for a word equation, make sure you use the correct chemical names. If you're asked for the symbol equation, always double-check that it's balanced correctly or you'll lose valuable marks.	22

Section Five — Acids, Bases and Reaction Rates

Rates of Reaction

1   The graph below shows the results of a rate of reaction experiment in which a student added the same mass of powdered marble to three different concentrations of hydrochloric acid. The same volume of acid was used each time and the acid was in excess.

a) Give the letter of the curve that shows the fastest rate of reaction. Give a reason for your answer.

....

....

....
                                                    [2]

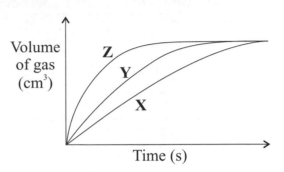

b) Explain why all three reactions produced the same volume of gas.

....

                                                    [1]

                                            [Total 3 marks]

2   A student mixes a measured mass of sodium carbonate with an excess of nitric acid in a flask. The student places the flask on a mass balance and records the mass of the reaction mixture every 10 seconds.

a) Describe what will happen to the mass of the reaction mixture as the reaction progresses.

....

                                                    [1]

b) A graph of the student's results is shown below.

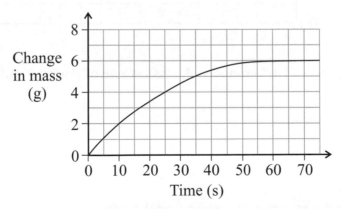

Calculate the rate of reaction (in g/s) for the period of time between 10 and 25 seconds.

Change in mass between 10 and 25 seconds = ... − .... = ... g

Rate = amount of reactant used ÷ time   = ... g ÷ .... s

                                             = ... g/s

                            Rate = ... g/s
                                                    [3]

                                            [Total 4 marks]

3   An experiment was set up to compare the rate of reaction of 5 g of magnesium ribbon with 20 ml of five different concentrations of hydrochloric acid. The volume of gas produced during the first minute of each reaction was recorded. The experiment was carried out twice for each concentration of acid. The results obtained are displayed in the table.

Concentration of HCl (mol/dm³)	Volume of gas produced (cm³)		
	Experiment 1	Experiment 2	Mean
2	92	96	94
1.5	63	65	....
1	44	47	....
0.5	20	50	20
0.25	9	9	9

a) Draw a ring around the anomalous result in the table.

[1]

b) i) Complete the table to show the mean volume of gas produced at each concentration.

[2]

   ii) Plot a graph of the results on the grid below. Label the *x* axis and draw a line of best fit.

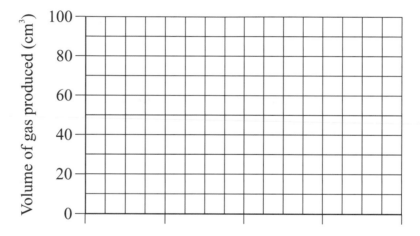

[4]

c) State which concentration of HCl produced the fastest rate of reaction and use the results to explain your answer.

....

....

[2]

d) The apparatus used in the experiment is shown on the right.
   i) What is the name of the piece of apparatus labelled X?

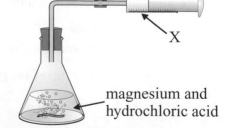

   ....

[1]

   ii) Name **one** other piece of apparatus needed for this experiment that is not shown in the diagram.

magnesium and hydrochloric acid

   ....

[1]   Score: ☐

[Total 11 marks]   **18**

Collision Theory and Catalysts

1 Reactions occur when particles collide with each other.

a) Explain, in terms of collisions, why increasing the pressure of the reactants in a gaseous reaction increases the reaction rate.

....

....

[2]

b) Reactions only happen when particles collide with sufficient energy.

i) What is the name given to the minimum amount of energy required?

....

[1]

ii) This energy is more likely to be reached when particles are travelling faster. Suggest how the speed of particles in a reaction could be increased.

....

[1]

[Total 4 marks]

2 A student investigated the reaction between 7 g of magnesium and excess sulfuric acid, using two different concentrations of acid. The student recorded the mass of the reaction mixture at the start and at every 10 seconds for 2 minutes, then calculated the change in mass for each reading.

a) On the axes below, sketch the curves you would expect the student to see for a high and a low concentration of acid. Label the curves.

[2]

Change in mass (g)

Time (s)

b) A teacher advised the student to think about hazards before conducting the experiment.

Identify **one** potential hazard of this experiment and suggest how the student could reduce the risk associated with this hazard.

....

....

....

[2]

[Total 4 marks]

Section Five — Acids, Bases and Reaction Rates

3   When sodium thiosulfate solution is mixed with hydrochloric acid, a precipitate forms. This causes the reaction mixture to go cloudy.

A student investigated the effect of temperature on the rate of this reaction. He mixed the reactants together in a flask and placed the flask on a piece of paper with a cross drawn on it. Then he timed how long it took for the mixture to become so cloudy that he could no longer see the cross.

a) Here are the results from the experiment:

Temperature (°C)	20	30	40	50	60
Time taken for cross to disappear (s)	201	177	145	112	82

i) Describe what happens to the rate of reaction as the temperature increases.

....

[1]

ii) Explain your answer to part a) i) in terms of particles and collisions.

....

....

....

....

[4]

b) Suggest **one** way in which the reliability of the results could be improved.

....

[1]

c) Next the student set up a similar experiment to investigate the effect of changing the concentration of the hydrochloric acid on the rate of the reaction.
The results of this experiment are shown in the table below.

Concentration of HCl (mol/dm³)	2.00	1.75	1.50	1.25	1.00
Time taken for cross to disappear (s)	13	23	38	50	67

i) What conclusion can be drawn from the results?

....

....

[1]

ii) Explain your conclusion in terms of collision theory.

....

....

[2]

[Total 9 marks]

4   A teacher demonstrated an experiment showing the effect of surface area on rate of reaction. The teacher added an excess of dilute hydrochloric acid to large marble chips and measured the volume of gas produced at regular time intervals. The teacher then repeated the experiment using the same mass of powdered marble. The teacher's results are shown on the graph below.

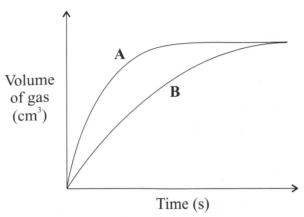

a) Give the letter of the curve, **A** or **B**, that shows the result of the experiment using powdered marble. Explain your choice in terms of particles and collisions.

....

....

....
[2]

b) On the graph, sketch the curve you would expect if the teacher repeated the experiment using small marble chips. Label it **C**.
[1]

c)   The teacher used the same mass of marble in each experiment.

    i) Give **two** other variables that the teacher should have controlled.

       1. ....

       2. ....
[2]

    ii) The teacher repeated the experiment using half the mass of large marble chips. All other conditions were kept the same.

       Suggest what effect you would expect this change to have on the total volume of gas produced.

       ....

       ....
[1]

d) Suggest **one** other method that could be used to measure the rate of this reaction.

....

....
[1]
[Total 7 marks]

Section Five — Acids, Bases and Reaction Rates

56

5 Hydrogen peroxide (H_2O_2) breaks down over time to give oxygen gas and water. Manganese dioxide (MnO_2) can act as a catalyst for this reaction.

a) Explain what a *catalyst* is.

....

....

....

[2]

b) A teacher measured out 20 cm³ of 1 mol/dm³ hydrogen peroxide solution into two flasks labelled **A** and **B**. She added some manganese dioxide to one flask. Then she recorded how much oxygen each flask gave off over a period of time. A graph of her results is shown below.

Give the letter of the curve that shows the flask with the catalyst added.
Explain your answer.

....

....

[2]

[Total 4 marks]

6 Catalysts can help to reduce the cost of industrial processes.

a) Suggest why catalysts reduce the cost of industrial processes.

....

....

....

[3]

b) Give **one** possible disadvantage of using a catalyst in an industrial process.

....

....

[1]

[Total 4 marks]

Score:

32

Crude Oil

1 Crude oil can be separated by fractional distillation into several different fractions.

a) Crude oil is a *mixture*. Explain what the term *mixture* means.

....

.... *[1]*

b) Crude oil is mostly made up of molecules containing only carbon and hydrogen atoms.

What name is given to these molecules?

....

[1]

c) The diagram on the right shows a fractionating column.

Which letter, **A-H**, shows where the fraction with the highest boiling point is tapped off?

....

[1]

d) *In this question you will be assessed on the quality of your English, the organisation of your ideas and your use of appropriate specialist vocabulary.*

Describe the process of separating crude oil using fractional distillation and explain how it works.

For this question you need to describe what's happening to the molecules in the oil at each step of the process. Remember to think about why the oil is heated to start with and why the fractionating column has a temperature gradient.

....

....

....

....

....

....

....

....

....

....

[6] **Score:**

[Total 9 marks]

9

Properties and Uses of Crude Oil

1   Crude oil contains alkanes.

a) i) The diagram below shows the structure of the first two alkanes. Complete the diagram by
        drawing the structure of the alkane with three carbon atoms and naming the molecule.

<u>**Alkane 1**</u>   <u>**Alkane 2**</u>   <u>**Alkane 3**</u>

$$
\begin{array}{c}
H \\
| \\
H-C-H \\
| \\
H
\end{array}
$$

methane

$$
\begin{array}{cc}
H & H \\
| & | \\
H-C-C-H \\
| & | \\
H & H
\end{array}
$$

ethane

....   *[2]*

ii) Which of these alkanes (**1-3**) has the lowest boiling point? Give a reason for your answer.

   ....

   ....
                                                                                                                      [1]

b) i) State the general formula of the alkanes. ....
                                                                                                                      [1]

   ii) Give the term for a group of compounds that can be represented by the same general formula.

   ....
                                                                                                                      [1]

c) Which **one** of these statements correctly describes alkane molecules? Tick **one** box.

   ☐   They are unsaturated — all their carbon-carbon bonds are single covalent bonds.

   ☐   They are saturated — all their carbon-carbon bonds are single covalent bonds.

   ☐   They are unsaturated — not all their carbon-carbon bonds are single covalent bonds.

   ☐   They are saturated — not all their carbon-carbon bonds are single covalent bonds.   *[1]*

                                                                                          [Total 6 marks]

2   Each hydrocarbon molecule in engine oil has a long chain of carbon atoms.

   Give **two** reasons why shorter chain hydrocarbons cannot be used in engine oil.

   1. ....

   ....

   2. ....

   ....
                                                                                                                      [Total 2 marks]

                                                                                          Score: ☐

                                                                                          8

Section Six — Crude Oil and Organic Chemistry

Environmental Problems

1   Hydrocarbons are often used as fuels. This can cause environmental problems.

a) Complete the sentences using words from the box.

| oxidised   released   reduction   combustion   combined   fusion |

When fuel is burned, energy is .... This is a .... reaction.

During this reaction, the carbon and hydrogen are ....

[3]

b) Complete the word equation for completely burning a hydrocarbon in air.

hydrocarbon + ... → .... + carbon dioxide

[2]

c) This graph shows the change in the percentage of carbon dioxide in the atmosphere since 1700.

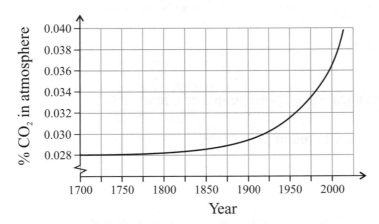

i) Describe the trend shown by the graph.

....

[1]

ii) What effect is the trend shown in the graph having on the Earth's climate?

....

[1]

d) Engines and power stations that burn fossil fuels can be adapted to use biofuels instead.

i) Give **two** advantages of replacing fossil fuels with biofuels.

1 ....

2 ....

[2]

ii) Give **two** disadvantages of replacing fossil fuels with biofuels.

1 ....

2 ....

[2]

[Total 11 marks]

2   Hydrogen can be burned as a fuel in some combustion engines.

a)   Give **one** other way that hydrogen can be used as a fuel.

....

[1]

b)   State the main advantage of using hydrogen fuel to power vehicles instead of hydrocarbon fuels.

....

....

[1]

c)   Give **two** reasons that hydrogen is not currently used on a large scale to power vehicles.

1. ....

....

2. ....

....

[2]

[Total 4 marks]

3   Ethanol made from sugar cane is an example of a renewable fuel.

a)   Which **two** elements are found in most fuels?

1. ....   2. ....

[2]

b)   Describe the process of producing pure ethanol from plant materials.

....

....

....

....

....

[3]

c)   A range of different plant materials can be used to produce fuels.

i)   What is the name for a type of fuel produced from plant material?

....

[1]

ii)   Name **one** fuel, other than ethanol, that is produced from plant materials.

....

[1]

[Total 7 marks]

Section Six — Crude Oil and Organic Chemistry

4   Exhaust fumes from cars, lorries and power stations are a major source of pollution.

a)   Exhaust fumes often contain carbon monoxide and solid particles.

i)   State **two** substances found in these solid particles.

1. ....

2. ....
[2]

ii)   Why are solid particles more likely to be formed in internal combustion engines than if the fuel was burnt in the open air?

....

....
[2]

iii) State **one** environmental problem caused by these solid particles.

....
[1]

b)   Nitrogen oxides are also produced in internal combustion engines.

i)   Under what condition do nitrogen oxides form?

....
[1]

ii)   Nitrogen oxides contribute to acid rain. Describe **one** effect of acid rain on the environment.

....

....
[1]

c)   Sulfur dioxide can also be found in smoke and exhaust fumes. It comes from sulfur in the fuels.

i)   Name an environmental problem caused by sulfur dioxide.

....
[1]

ii)   Complete the sentences below.

Waste gases from .... can be sprayed

with a slurry of limestone to remove the sulfur dioxide.

Liquid fuel can be treated with hydrogen to remove the sulfur before it is burnt — this is used

to reduce the sulfur dioxide emissions of ....
[2]
[Total 10 marks]

Score: ☐

32

Cracking Crude Oil

1 Cracking alters the molecules obtained in fractional distillation.

a) *Cracking* is a necessary process in the oil industry.

i) Describe what is meant by *cracking*.

....

....

[2]

ii) Complete the sentence below about the products of cracking.

The main products of cracking are saturated hydrocarbons called ....

and unsaturated hydrocarbons called ....

[2]

iii) Give **one** use of a product of cracking.

....

[1]

b) The apparatus shown below can be used to crack paraffin in the lab.

i) Suggest what role the silica is playing.

....

....

[1]

ii) Name two products of this experiment
that will collect in the gas jar.

1. ....

2. ....

[2]

mineral wool
soaked in
paraffin

gas jar

silica

iii) Give **one** potential hazard of this experiment and suggest a suitable safety precaution.

....

....

[2]

c) Give another method of cracking paraffin.

....

[1]

d) What type of reaction is cracking?

....

[1]

[Total 12 marks]

Score:

12

Alkenes and Ethanol

1 Alkenes are a group of unsaturated hydrocarbon molecules.

a) State the general formula of the alkenes. ....
[1]

b) Complete the table to show the missing information for the two alkenes given.

Name of alkene	Formula	Displayed structure
ethene	....	
....	C_3H_6	

[4]

[Total 5 marks]

2 A student tested five different organic compounds using bromine water.

a) Diagrams **A-E** show the displayed structures of the compounds that the student tested.

A   **B**   **C**   **D**   **E**

i) Give the letters of **two** unsaturated compounds. .... and ....
[1]

ii) Explain how you know that they are unsaturated compounds.

....

....
[1]

b) The student added each of the compounds to separate tubes of bromine water and shook them.

State the colour change (if any) that occurred when compounds **C** and **E** were shaken
with bromine water.

....

....

....

....
[2]

[Total 4 marks]

3 Two different methods can be used to manufacture ethanol.
The incomplete table below shows some information about the two methods.

a) Complete the table.

Method	Reaction	Reaction conditions	Problems
A	$C_2H_4 + \text{.................} \rightarrow C_2H_5OH$	300 °C, 60 - 70 atmospheres, phosphoric acid	Expensive equipment
B	$C_6H_{12}O_6 \rightarrow 2CO_2 + 2C_2H_5OH$	20 - 35 °C	Labour-intensive

[1]

b) One reactant in method **A** is a hydrocarbon made by cracking crude oil.

i) Name the reactant. ....
[1]

ii) Name the homologous series the reactant belongs to. ....
[1]

c) Suggest the role of the phosphoric acid in method **A**.

....
[1]

d) A particular country has a good supply of crude oil. The workers in this country are paid very high wages. Suggest which method, **A** or **B**, would be most suitable for manufacturing ethanol in this country. Give a reason for your answer.

....

....

....
[1]
[Total 5 marks]

4 A known volume of ethanol was burned to heat some water.
The diagrams below show the thermometer that was used to measure the temperature of the water (in °C) before and after the ethanol was burned.

Before   After

20   40

10   30

Use the diagrams to work out the temperature change of the water.

.... °C
[Total 2 marks]

Score:

16

Polymers

1 Poly(ethene) is a polymer made from ethene.

a) i) Complete the equation below to show the polymerisation of ethene to form poly(ethene).

$$n\begin{pmatrix} H & H \\ | & | \\ C=C \\ | & | \\ H & H \end{pmatrix} \longrightarrow$$

[3]

ii) State what is meant by *polymerisation*.

....
[1]

b) A plastics company has samples of two different polymers.
Polymer **1** is a low density poly(ethene) (LDP). Polymer **2** is a high density poly(ethene) (HDP).

i) The company performs some tests to investigate the properties of the polymers.
The table below shows their results.

Polymer	Type of polyethene	Density	Softening temperature	Flexibility
1	LDP	Low	90 °C	High
2	HDP	High	120 °C	Fairly low

The company wants to make freezer bags. Which of the two polymers should they use?
Give a reason for your choice.

....
....
[1]

ii) Which of the diagrams below, **A-C**, represents the structure of LDP?

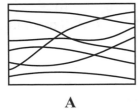 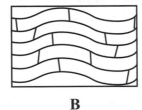 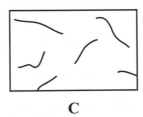

A **B** **C**

Diagram ....
[1]

iii) HDP and LDP are examples of different polymers that are made from the same monomer.
How can polymers with different properties can be made from the same monomer?

....
....
[1]

[Total 7 marks]

2   Most polymers made from crude oil are difficult to dispose of, so it's good for the environment to reuse them as many times as possible. Thermosoftening polymers are easier to recycle than thermosetting polymers.

a) Explain why most polymers are difficult to dispose of.

....

....

[2]

b) New plastics have been developed to make carrier bags that are easier to dispose of. Draw a ring around the substance in the plastic that makes them easy to dispose of.

   poly(ethene)   **cane sugar**   **cornstarch**   **hydrogel**

[1]

c) Describe the bonds between the polymer chains in thermosoftening and thermosetting plastics and explain how they affect the melting points of both types of plastic.

....

....

....

....

....

....

[4]

[Total 7 marks]

3   A manufacturer tested a new polymer for making pillows. The results are shown in the graph on the right.

a) Describe how the pressure needed to squash the polymer is related to temperature.

....

....

....

[2]

b) Suggest why this polymer would be suitable for making pillows.

....

....

[1]

[Total 3 marks]

Score: ☐

17

Section Six — Crude Oil and Organic Chemistry

Alcohols

1 Alcohols are a group of chemicals with many uses.

a) The incomplete table below shows details of three alcohols. Complete the table.

Name of alcohol	Number of carbon atoms	Molecular formula	Displayed structure
Methanol	....	....	

....	2	....	H H
			  │ │
			H—C—C—O—H
			  │ │
			H H

Propanol	3	C_3H_7OH	

[6]

b) What is the functional group found in all alcohols? ....
[1]

c) Complete the sentences below about alcohols.

   i) A solution of propanol in water has a pH of ....
[1]

   ii) .... is the main alcohol used in alcoholic drinks.

      Alcohols can also be used as .... and ....
[3]

d) Name the two products that are formed when an alcohol is burned in air.

   1. ....   2. ....
[2]

e) Draw a ring around each of the substances below that are alcohols.

    butanol   **C_3H_8**   **ethene**   **$C_5H_{11}OH$**   **ethyl ethanoate**   **C_3H_7COOH**
[2]

f) Complete the equation below, which shows an alcohol reacting with sodium.

    $2C_3H_7OH + 2Na \rightarrow 2C_3H_7ONa +$ ....
[1]

[Total 16 marks]

Score: ☐
16

 ☐   ☐   ☐

Carboxylic Acids

1   Ethanoic acid is a carboxylic acid.

a) Give the functional group of carboxylic acids. ....
[1]

b) What is the displayed structure of ethanoic acid? Draw a ring around the correct answer.

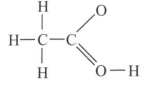   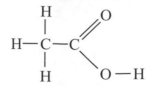   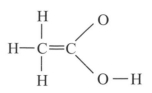
[1]

c) Complete the following sentences.

.... can be oxidised to give ethanoic acid. This can be done using either

.... or oxidising agents. Ethanoic acid is the main acid in ....
[3]

d) When ethanoic acid is dissolved in water it forms a weak acidic solution.

i) Explain why this is.

....

....

....
[2]

ii) Hydrochloric acid is a strong acid. How would you expect the pH of a solution of hydrochloric acid to compare to the pH of a solution of ethanoic acid of the same concentration?

....
[1]

[Total 8 marks]

2   Propanoic acid is useful as a preservative and for making other chemicals.

a) Propanoic acid reacts with alcohol in the presence of an acid catalyst.
What type of compound forms?

....
[1]

b) Complete and balance the equation for the reaction of propanoic acid with potassium carbonate.

.... + K_2CO_3 → .... + ... + $2CH_3CH_2COOK$
[2]

[Total 3 marks]

Score: ☐

11

Esters

1 When ethanol is heated with ethanoic acid in the presence of a catalyst, an ester is produced.

a) Name a catalyst that could be used for this reaction.

....
[1]

b) Complete the chemical equation for this reaction.

CH_3COOH + ... $\rightarrow$ .... + H_2O
[1]

c) Name the ester produced in this reaction.

....
[1]

d) State the functional group of the esters.

....
[1]

e) Some students want to carry out the reaction to produce an ester. Their teacher reminds them that alcohols and esters are both flammable and should not be heated over an open flame.

Suggest a method that the students could use to heat the reaction mixture safely.

....

....
[1]
[Total 5 marks]

2 Esters are volatile compounds that are often used in perfumes.

a) Explain what is meant by the term *volatile*.

....
[1]

b) Apart from volatility, give **one** other property of esters that makes them suitable for use in perfumes.

....
[1]

c) Other than perfumes, give **one** use of esters in the chemical industry.

....
[1]
[Total 3 marks]

Score: [   ]

8

Energy Transfer in Reactions

1   Use words from the box to complete the following passage about energy transfer in reactions. Each option may be used once, more than once or not at all.

> less   exothermic   greater   endothermic

Breaking bonds is an .... process and forming bonds is

an .... process. In an exothermic reaction, the energy

released during bond formation is .... than the energy

used to break the old bonds.

[Total 3 marks]

2   During the following reaction, the temperature of the reaction mixture decreases.

$$AB + C \rightarrow AC + B$$

a) State, with a reason, whether the reaction is exothermic or endothermic.

....

....

[1]

b) Which bond is stronger, A–B or A–C? Explain your answer.

....

....

[1]

c) i) What is meant by the *enthalpy change* of a reaction?

....

[1]

ii) State whether the enthalpy change of the above reaction is positive or negative.

....

[1]

iii) Give the symbol that is used to represent enthalpy change.

....

[1]

[Total 5 marks]

Score: ☐

8

Energy Transfers and Reversible Reactions

1 Suggest possible everyday applications for each of the following reactions.

a) The exothermic oxidation of zinc metal by copper sulfate solution.

....
[1]

b) The endothermic process of dissolving ammonium nitrate in water.

....
[1]
[Total 2 marks]

2 Limestone ($CaCO_3$) breaks down when heated to form calcium oxide and carbon dioxide.

a) What type of reaction is this? ....
[1]

b) State whether this reaction is exothermic or endothermic, and explain your answer.

....

....
[1]
[Total 2 marks]

3 When hydrated copper(II) sulfate is heated, anhydrous copper(II) sulfate and steam are produced.

$$CuSO_4.5H_2O_{(s)} \rightleftharpoons CuSO_{4(s)} + 5H_2O_{(g)}$$

a) The enthalpy change for the forward reaction is –78.2 kJ/mol.
State the enthalpy change of the reverse reaction.

.... kJ/mol
[1]

b) A student has a beaker containing some anhydrous copper(II) sulfate powder.
A few drops of water are added to the beaker from a pipette.

i) Describe what happens to the colour of the copper(II) sulfate.

....
[1]

ii) What happens to the temperature of the mixture in the beaker?

....
[1]
[Total 3 marks]

Score:

7

Section Seven — Energy and Equilibria

Energy Level Diagrams

1   Chemical reactions involve enthalpy changes.

a) What is meant by the term *activation energy*?

....

[1]

b) Explain how catalysts increase the rate of a reaction.

....

[1]

c) The diagrams below represent the energy changes in four different chemical reactions.

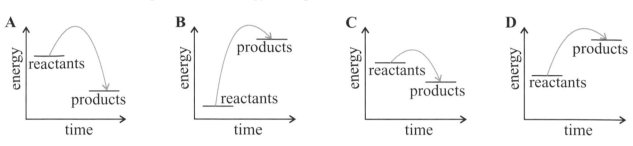

Write the letter (**A**, **B**, **C** or **D**) of:

i) any **one** of these reactions that is exothermic.   .... *[1]*

ii) any **one** of these reactions that is endothermic.   .... *[1]*

iii)the reaction with the smallest change in energy.   .... *[1]*

iv)the reaction with the largest activation energy.   .... *[1]*

d) An energy level diagram is shown on the right.

i) Give the enthalpy change for this reaction.

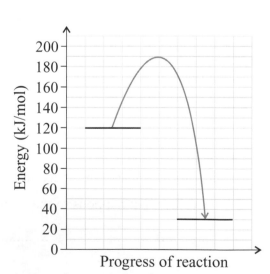

....

[1]

ii) Add a label to the graph to show the
overall energy change of the reaction.   *[1]*

iii)Give the activation energy of this reaction.

....

[1]

iv)Sketch the curve that you would expect to see if the reaction was repeated with a catalyst.   *[1]*

[Total 10 marks]

Exam Practice Tip   **Score**

In an exam you might be given some data and asked to draw an energy level diagram, so get plenty of practice
doing them — for both exothermic and endothermic reactions. As well as graphs, examiners are also really keen
on neatness (no surprises there), so don't forget to take a ruler and a nice sharp pencil into the exam with you.

9

Bond Dissociation Energy

1 Calculate the enthalpy change for the combustion of methane. Use the equation and the bond dissociation energies given below.

$$CH_4 + 2O_2 \rightarrow CO_2 + 2H_2O$$

C–H = +412 kJ/mol   O=O = +498 kJ/mol   C=O = +743 kJ/mol   O–H = +463 kJ/mol

Total energy absorbed to break the bonds in the reactants:

(4 x C–H) + (2 x O=O) = (4 x ....) + (2 x ....) = ...

Total energy released when the bonds in the products are formed:

(2 x C=O) + (4 x O–H) = (2 x ....) + (4 x ....) = ...

Enthalpy change = total energy absorbed to break bonds – total energy released in making bonds

= .... – ....

Enthalpy change = ... kJ/mol

[Total 3 marks]

2 Calculate the enthalpy change for the combustion of hydrazine, N_2H_2. Use the equation and the bond dissociation energies given below.

$$N_2H_4 + O_2 \rightarrow N_2 + 2H_2O$$

N–N = +158 kJ/mol   N–H = +391 kJ/mol   O=O = +498 kJ/mol
N≡N = +945 kJ/mol   O–H = +463 kJ/mol

Enthalpy change = ... kJ/mol

[Total 3 marks]

Score: ☐

6

Measuring Energy Transfer

1   A student investigated the temperature change of a reaction using the apparatus shown below. The student dissolved some calcium chloride in water, and measured the temperature of the reaction over the first 30 seconds.

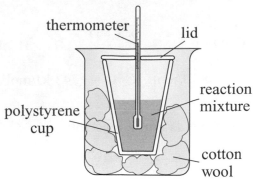

a) State the purpose of:

   i) the cotton wool ....

   [1]

   ii) the lid ....

   [1]

b) The student's results are plotted on the graph shown on the right.

   What was the temperature increase during the reaction?

   .... °C

   [1]

c) i) Suggest why it is difficult to get an accurate result for the energy change in an experiment like this.

   ....

   ....

   ....

   [1]

   ii) How could the reliability of the results be improved?

   ....

   ....

   [1]

d) Name **one** other type of reaction that this method could be used to study.

   ....

   [1]

   [Total 6 marks]

2 A scientist conducted a calorimetry experiment to measure the energy produced when petrol is burnt. 0.7 g of petrol was burnt in a spirit burner placed underneath a copper can containing 50 g of water. The temperature of the water increased by 30.5 °C.

a) Why was a copper can chosen to hold the water?

....

[1]

b) It takes 4.2 J to raise the temperature of 1 g of water by 1 °C.
Calculate the heat energy change in the experiment using the formula $Q = m \times c \times \Delta T$

Remember: Q is the heat energy change (in J), m is the mass of water used (in g),
c is the specific heat capacity of water (4.2) and ΔT is the temperature change (in °C).

Heat energy change = ... J

[1]

c) Use your answer to b) to calculate the energy released per gram of petrol.
Give your answer in kJ/g.

Energy released = ... kJ/g

[2]

[Total 4 marks]

3 In a calorimetry experiment, a student found that burning 1.15 g of ethanol raised the temperature of 50 g of water by 34.5 °C. She calculated that this was a heat energy change of 7245 J.

a) Calculate the number of moles of ethanol that
the student burnt in her experiment.

To work out the number of moles, you first need to know the relative formula mass.

M_r of ethanol = ...

So number of moles = ... ÷ .... = ...

Moles = ... mol

[2]

b) Calculate the molar enthalpy change (in kJ/mol) for the combustion of ethanol.

Watch out for the units here...

Molar enthalpy change = ... kJ/mol

[2]

[Total 4 marks]

Exam Practice Tip

Calculations, calculations, calculations. Everyone's favourite. You'll have heard it a thousand times before, but always <u>check your working</u> after you've done a question, and check the <u>units</u> too. For a molar enthalpy question you also need to think about <u>signs</u>: '-' for exothermic reactions and '+' for endothermic reactions.

Score

[   ]

14

Section Seven — Energy and Equilibria

76

Equilibrium and Yield

1   In the reaction below, substances **A** and **B** react to form substances **C** and **D**.

$$2A\,(g) + B\,(g) \rightleftharpoons 2C\,(g) + D\,(g)$$

a) What can you deduce about this reaction from the symbol $\rightleftharpoons$?

....

[1]

b) In the above reaction, the forward reaction is exothermic.

   i) Does the reverse reaction take in or give out heat energy? Explain your answer.

....

....

[2]

   ii) State whether lowering the temperature will increase the yield of the forward reaction
      or the reverse reaction. Explain your answer.

....

....

....

[1]

   iii) Explain why changing the temperature of a reversible reaction always affects
      the position of the equilibrium.

....

....

....

[2]

c) State and explain the effect of changing the pressure on the position of equilibrium
   in the above reaction.

....

....

....

[2]

d) Ammonium chloride is heated in an open flask inside a fume cupboard to form ammonia gas and
   hydrogen chloride gas. Why can this reaction not reach equilibrium even though it is reversible?

....

....

[1]

[Total 9 marks]

Section Seven — Energy and Equilibria

2 Calcium carbonate will decompose when it is heated to a high temperature. This is a reversible reaction.

$$CaCO_3 \rightleftharpoons CaO + CO_2$$

a) Identify the product(s) of the reverse reaction.

....

[1]

b) If this reaction is carried out in a closed system it will reach equilibrium.
Complete the following sentence.

At equilibrium, the forward and backward reactions happen at .... rate.

[1]

[Total 2 marks]

3 When ammonium chloride is heated, it breaks down into ammonia and hydrogen chloride. The reaction is reversible.

$$NH_4Cl\,(s) \rightleftharpoons NH_3\,(g) + HCl\,(g)$$

Two students are trying to deduce the optimum conditions to favour the forward reaction.

The first student suggests a temperature of 375 °C and a pressure of 1 atmosphere.
The second student suggests a temperature of 250 °C and a pressure of 5 atmospheres.

Using your knowledge of reversible reactions, state which set of conditions would be more favourable for the forward reaction. Explain your answer.

....

....

....

....

....

....

....

....

....

....

....

[Total 5 marks]

Score:

16

The Haber Process

1 The Haber process is used to make ammonia, which is widely used in industry.

a) Complete the chemical equation for the reaction.

.... + ... $\rightleftharpoons$ $2NH_3$

[2]

b) State the name of the two reactants in the forward reaction and a source of each one.

Reactant 1: .... Source: ....

Reactant 2: .... Source: ....

[4]

[Total 6 marks]

2 The Haber process is carried out under a specific set of conditions.

a) i) Tick one box to show the pressure used for the Haber process.

☐ 1000 atmospheres ☐ 200 atmospheres ☐ 450 atmospheres

[1]

ii) The temperature used in the Haber process is 450 °C. Explain why this temperature is chosen.

....

....

....

....

....

....

[4]

b) Name the catalyst used in the Haber process.

....

[1]

c) In the Haber process, gases pass through a reaction chamber and then enter a cooling chamber.

Explain how cooling the gases allows ammonia to be separated from
unused hydrogen and nitrogen, and state what happens to these unused gases.

....

....

....

[3]

[Total 9 marks]

Score: ☐

15

Electrolysis

1 Lead bromide is an ionic compound. It can be broken up into its elements using electrolysis. This process is shown in the diagram below.

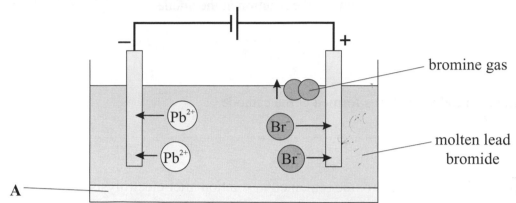

a) Name the substance labelled **A** in the diagram.

....

[1]

b) In the diagram above, molten lead bromide is the electrolyte.
   Explain why an ionic compound must be molten or dissolved for electrolysis to work.

....

....

....

[3]

c) i) Complete the balanced half-equations for the processes
   that occur during the electrolysis of molten lead bromide.

   Cathode: Pb^{2+} + ... → Pb

   [3]

   Anode :   .... → Br_2 + ...

ii) At which electrode does reduction occur? Explain your answer.

....

....

[2]

iii) Explain why the bromide ions move to the anode during electrolysis.

....

....

[2]

[Total 11 marks]

Score: ☐

11

 ☐   ☐ ☺ ☐

Electrolysis of Sodium Chloride Solution

1 A highly concentrated aqueous solution of sodium chloride was electrolysed.

a) Chlorine gas was formed at the anode.
Write a balanced half equation for the reaction at the anode.

....

[2]

b) Name the product that was formed at the cathode.

....

[1]

c) Explain why sodium hydroxide remained in solution after electrolysis was complete.

....

....

....

[3]

d) Electrolysis was carried out on a dilute solution of sodium chloride.
Instead of chlorine, oxygen gas was formed at the anode.

> Think about the factors
> that affect what product
> gets made at the anode.

i) Explain why oxygen gas is only formed at the anode when
a dilute solution of sodium chloride solution is used.

....

....

....

....

[4]

ii) Name the product formed at the cathode when a dilute solution of sodium chloride is used.

....

[1]

e) The products of the electrolysis of aqueous sodium chloride are useful in industry.
Complete the sentences.

i) Sodium hydroxide is a strong alkali that is used to make ....

ii) Chlorine is used in the production of ....

[2]

[Total 13 marks]

Score:

13

Section Eight — Electrolysis and Analysis

Electrolysis of Aluminium and Electroplating

1   Electrolysis can be used to extract some metals from their ores, but is an expensive process.

a)   What makes electrolysis so expensive?

....

[1]

b)   Aluminium is extracted from purified aluminium oxide by electrolysis.

i)   How does dissolving aluminium oxide in cryolite reduce the cost of extracting aluminium?

....

....

[2]

ii) A different reaction occurs at each electrode in the electrolysis of aluminium oxide.
Give the two ionic half-equations that represent these reactions.

....

....

[3]

iii)Explain why carbon dioxide is produced as a by-product of this process.

....

....

....

[2]

[Total 8 marks]

2   A factory uses electroplating to cover nickel alloy forks in a layer of silver.

a)   Give the name of **one** ion that the electrolyte must contain.

....

[1]

b)   Suggest **one** reason why the company might want to electroplate a nickel alloy fork with silver.

....

....

[1]

c)   Give **one** other reason why an object might be electroplated.

....

[1]

[Total 3 marks]

Score:

11

Tests for Positive and Negative Ions

1 Potassium chloride is used to replace some of the sodium chloride in low-sodium table salt. A flame test can be used to tell the difference between potassium chloride and sodium chloride.

a) Describe how to carry out a flame test.

....

....

....

[2]

b) Explain how you could tell from a flame test that a substance was potassium chloride and not sodium chloride.

....

....

[2]

c) Potassium sodium tartrate is another food additive that contains potassium. Suggest why you could not use a flame test to show that this compound contains potassium.

....

....

[2]

[Total 6 marks]

2 A student adds a few drops of NaOH solution to solutions of different metal compounds.

a) Complete her table of results, showing the metal ions and the colours of the precipitates that form.

Metal ion in solution	Colour of precipitate formed
Fe^{2+}	....
....	blue
Fe^{3+}	....

[3]

b) Complete the balanced ionic equation for the reaction of iron(II) ions with hydroxide ions by inserting state symbols.

$$Fe^{2+}(\text{.........}) + 2OH^-(\text{.........}) \rightarrow Fe(OH)_2(\text{.........})$$

[1]

[Total 4 marks]

3 A student has a sample of an ionic compound and wants to find out what negative ions it contains.

a) Give the chemical formula and charge of the negative ions present in the following compounds.

i) barium sulfate ....
[1]

ii) potassium iodide ....
[1]

iii)magnesium carbonate ....
[1]

b) The student wants to test the compound for the presence of sulfate ions.

i) State which **two** reactants are used to test for sulfate ions.

Reactant 1 ....

Reactant 2 ....
[2]

ii) What would be observed after adding these reactants to a solution of a sulfate compound?

....
[1]

c) The student tested the compound to see if it contained carbonate ions. The student added a solution to the compound, collected the gas that it gave off and bubbled the gas through limewater.

i) Name the solution that the student added. ....
[1]

ii) The compound is a carbonate. What gas did it give off? ....
[1]

d) The student is given a solution of another compound and told that it contains either chloride or bromide ions. Describe a test the student could perform to discover which of these ions it contains.

....

....

....

....

....

....
[4]

e) Complete the following symbol equations for reactions involved in negative ion tests.

i) Ag^+ + ... $\rightarrow$ AgCl   ii) Ba^{2+} + ... $\rightarrow$ $BaSO_4$
[1]   [1]

[Total 14 marks]

Section Eight — Electrolysis and Analysis

84

4 A student analysed an unknown ionic compound in order to discover what elements it contained.

a) The student added a few drops of sodium hydroxide solution to an aqueous solution of the unknown compound. A white precipitate formed immediately.

i) The student said this showed that the compound must contain aluminium ions. Explain why the student is incorrect.

....

[1]

ii) The student continued to add sodium hydroxide, but no change was observed. What does this observation tell the student about the compound? Explain your answer.

....

....

[2]

b) The student performed a flame test on the unknown compound. He wiped a wire loop on a paper towel to clean it. Then he dipped the wire loop into the compound, placed it in a yellow Bunsen burner flame and observed the colour.

i) Identify **one** error that the student made and suggest how he could correct it.

....

....

[2]

ii) Suggest **one** safety precaution the student should take when performing a flame test.

....

....

[1]

iii) The student repeated the flame test correctly and observed that the unknown compound burned with the same colour of flame as calcium carbonate. What colour was the flame?

....

[1]

c) The student added dilute nitric acid followed by silver nitrate solution to a solution of the unknown compound. A yellow precipitate formed.

Give the name of the unknown compound.

....

[1]

[Total 8 marks]

Exam Practice Tip

The bad news here is that you need to know all the ion tests and their results off by heart — they're always cropping up in exam questions. The good news is that most questions on this topic have the same format: 1) spot which ion test is being used, and 2) use the results to work out what ions a substance contains.

Score

32

Section Eight — Electrolysis and Analysis

Separating Mixtures

1 Water and propanone form a mixture when combined.

a) Complete the sentences.

A mixture is made up of two or more .... or compounds that are not

chemically bonded to each other.

Unlike a compound, a mixture can be separated by .... methods.

[2]

b) The distillation apparatus shown below was used to separate a mixture of propanone and water. The boiling point of propanone is 56 °C.

Use words from the box to complete the table.
Each option may be used once, more than once, or not at all.

| **no liquid**   **water**   **propanone**   **both liquids** |

Think about what temperature the thermometer would have to show before each liquid boiled and escaped from the flask.

Temperature on thermometer	Contents of the beaker
30 °C	....
65 °C	....
110 °C	....

thermometer

condenser

flask

HEAT

beaker

[2]

[Total 4 marks]

---

2 Lawn sand is a mixture of insoluble sharp sand and soluble ammonium sulfate fertiliser.

a) Describe how you would obtain pure, dry samples of the two components of lawn sand in the lab.

....

....

....

....

....

[5]

b) A student separated 51.4 g of lawn sand into sharp sand and ammonium sulfate.
After separation, the total mass of the two products was 52.6 g. Suggest **one** reason for this error.

....

....

[1]

[Total 6 marks]

Score: ☐

10

Paper Chromatography

1   A scientist uses paper chromatography to compare different inks.

a) Describe the method used to set up a paper chromatography experiment to compare the inks.

....

....

....

....

....

....

[3]

b) The scientist uses paper chromatography to compare an ink used on a document (**X**) with the ink in three different printers (**A**, **B** and **C**). The chromatogram is shown on the right.

Which printers could **not** have produced the document?

....

[1]

c) Explain why the chemicals in the inks separated out.

....

....

....

....

[2]

°**Y**

X   A   B   C

d) Spots on a chromatogram can be compared using their R_f values.

i) Use the diagram above to fill in the missing values. Remember to include units.

Distance travelled by solvent: ....

Distance travelled by chemical **Y**: ....

[2]

ii) Use your answers to part d) i) to calculate the R_f value for chemical **Y**.

R_f value = ...

[1]

[Total 9 marks]

Score: ☐

9

Instrumental Methods

1   Forensic scientists use instrumental methods to analyse substances found at crime scenes.

a) Give **three** advantages of using instrumental methods to analyse substances.

1. ....

2. ....

3. ....

[3]

b) *In this question you will be assessed on the quality of your English,
the organisation of your ideas and your use of appropriate specialist vocabulary.*

A scientist uses gas chromatography linked to mass spectroscopy (GC-MS) to analyse
a sample of a substance found at a crime scene.

Describe how GC-MS can be used to separate and identify the different chemical compounds
that make up the substance.

For this question, start off by describing how gas chromatography separates out the compounds
that make up a substance. Then explain how mass spectroscopy could be used to identify them.

....

....

....

....

....

....

....

....

....

....

....

....

[6]

[Total 9 marks]

Exam Practice Tip

Don't be intimidated by a big six-marker like this — read the question carefully and take a moment to think
about what you're going to write. Your answer has to be in full sentences and laid out in a way that makes
sense. You'll get marks for using specialist vocab too — 'molecular ion' would be an example for this question.

Score

9

Candidate Surname	Candidate Forename(s)

Centre Number	Candidate Number

Level 1/2 Certificate in Chemistry
Paper 1

Practice Paper
Time allowed: 90 minutes

You must have:
• A ruler.
• A calculator.
• A periodic table.

Total marks:

Instructions to candidates
• Use **black** ink to write your answers.
• Write your name and other details in the spaces provided above.
• Answer **all** questions in the spaces provided.
• In calculations, show clearly how you worked out your answers.

Information for candidates
• The marks available are given in brackets at the end of each question.
• There are 90 marks available for this paper.
• You should answer Question 7 (b) with continuous prose.
  You will be assessed on the quality of your English, the organisation of your ideas and your use of appropriate specialist vocabulary.

Advice for candidates
• Read all the questions carefully.
• Write your answers as clearly and neatly as possible.
• Keep in mind how much time you have left.

Answer **all** questions

1   Atoms contain protons, neutrons and electrons.

1 (a)   Complete the table to show the relative charges and relative masses of protons, neutrons and electrons.

Particle	Relative mass	Relative charge
Proton	....	....
Neutron	....	....
Electron	very small	−1

[2]

1 (b)   The table below shows the numbers of protons, neutrons and electrons in six different atoms.

Atom	Number of protons	Number of neutrons	Number of electrons
A	5	6	5
B	7	7	7
C	6	8	6
D	6	6	6
E	10	10	10
F	4	5	4

Which **two** atoms are isotopes of the same element? Explain your answer.

Atoms .... and ....

Explanation ....

....

[2]

Question 1 continues on the next page

Turn over ▶

Practice Paper 1

1 (c) Zinc appears in the periodic table as shown below.

65
Zn
Zinc
30

How many protons, neutrons and electrons are there in an atom of zinc?

Protons: ....

Neutrons: ....

Electrons: ....

[3]

1 (d) Zinc sulfate is a *compound* with the formula $ZnSO_4$.

1 (d) (i) What is a *compound*?

....

....

[2]

1 (d) (ii) Calculate the relative formula mass (M_r) of zinc sulfate.

....

....

....

Relative formula mass = ...

[2]

[Total 11 marks]

2 A student was investigating the properties of the alkanes.

2 (a) The student used some data to draw the graph below, showing the number of carbon atoms in alkane molecules plotted against their boiling points. Pentane was missing from the data.

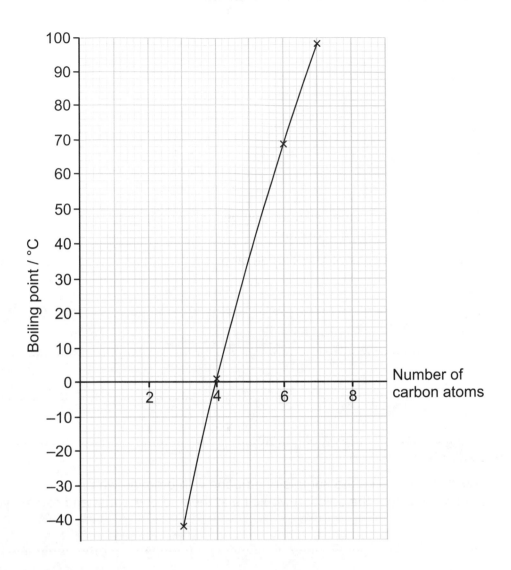

2 (a) (i) Use the graph to estimate the boiling point of pentane, C_5H_{12}.

.... °C

[1]

2 (a) (ii) Describe what the graph shows about the size of alkane molecules and their boiling points.

....

....

[1]

Question 2 continues on the next page

Turn over ▶

2 (b) A teacher gives the student a sample of propane, C_3H_8.

2 (b) (i) Calculate the percentage mass of carbon in propane.

Clearly show how you work out your answer.

....

....

....

Percentage mass of carbon = ... %

[2]

2 (b) (ii) Propane and other small alkanes can be produced by cracking larger hydrocarbons, like decane.

Describe the process of cracking decane.

....

....

....

....

[2]

2 (b) (iii) The student shakes the sample of propane with bromine water.

State what colour the solution would be at the end of the test and explain your answer.

....

....

....

[2]

[Total 8 marks]

3 Iron and its alloys can be used to make pipes.

3 (a) Iron can be extracted from iron ore using a blast furnace.

3 (a) (i) Most of the iron produced in blast furnaces is converted into steel.
Describe how the composition of steel differs from blast furnace iron.

....

....
[2]

3 (a) (ii) Iron can also be obtained by recycling old metal items.
Suggest **two** reasons why it is important that metals are recycled.

1. ....

....

2. ....

....
[2]

3 (b) Iron pipes need protection from rusting.

3 (b) (i) State the conditions that are needed for iron to rust.

....

....[2]

3 (b) (ii) Complete the sentence

When iron rusts it is .... — this means that it gains oxygen.
[1]

Question 3 continues on the next page

Turn over ▶

3 (c) Paint can be used to stop the outside of iron pipes rusting, but they will eventually need repainting or replacing. An alternative to this is to connect a large piece of magnesium to the pipe, as shown in the diagram below.

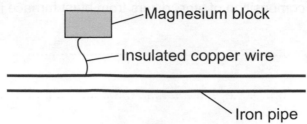

—Magnesium block

—Insulated copper wire

Iron pipe

3 (c) (i) What name is given to this kind of corrosion protection?

....

[1]

3 (c) (ii) Explain how the magnesium protects the iron from rusting.

....

....

....

[2]

[Total 10 marks]

4 All alcohol molecules contain the functional group –OH.
The displayed structures of two alcohols are shown below.

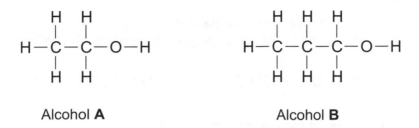

Alcohol **A**   Alcohol **B**

4 (a) Alcohol **A** is ethanol.

4 (a) (i) Ethanol is an alcohol that can be used instead of petrol as fuel in a car engine.
Give **one** other use of ethanol.

....

[1]

4 (a) (ii) Complete the sentences.

When a sample of pure ethanol is added to water it ....

to form a colourless solution with a pH of ....

[2]

Question 4 continues on the next page

Turn over ▶

4 (b) Name alcohol **B**.

....
[1]

4 (c) Alcohol **B** can be used to make propyl ethanoate.
The symbol equation for the formation of propyl ethanoate is shown below.

$$C_3H_7OH \ + \ CH_3COOH \ \rightarrow \ CH_3COOC_3H_7 \ + \ H_2O$$

4 (c) (i) Give the chemical name of the compound CH_3COOH.

....
[1]

4 (c) (ii) Give the homologous series and the functional group of propyl ethanoate.

Homologous series: ....

Functional group: ....
[2]

4 (c) (iii) Propyl ethanoate is used in some perfumes because it has a pleasant, fruity smell.

Give **one** other property of propyl ethanoate that makes it suitable for use in perfumes.

....
[1]
[Total 8 marks]

5 A student is given solutions of four ionic compounds that contain either chlorine or iodine.

5 (a) (i) Complete the diagram below to show the electronic structure of a chlorine atom.
Mark each electron using an 'X'.

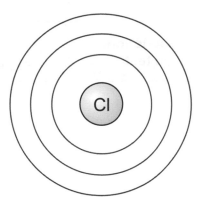

[1]

5 (a) (ii) Explain how the group number of chlorine in the periodic table
is related to its electronic structure.

....

....

....

[2]

5 (b) The student bubbles chlorine through sodium iodide solution.

Describe the chemical change that takes place and explain why it happens
in terms of the relative positions of chlorine and iodine in the periodic table.

....

....

....

....

....

[3]

Question 5 continues on the next page

Turn over ▶

5 (c) The labels have fallen off the student's bottles of solutions.
The student is no longer sure which solution is which.

Suggest tests that could be used to distinguish between the following pairs of compounds in solution. You should describe the tests and the results expected for each solution.

Solution A	Solution B	Description of test	Observations	
			Solution A	Solution B
iron(II) chloride	iron(III) chloride	....	....	....

Solution C	Solution D	Description of test	Observations	
			Solution C	Solution D
sodium chloride	sodium iodide	....	....	....

[4]

[Total 10 marks]

6   The table gives data for some physical properties of a selection of substances.

Substance	Melting point	Boiling point	Electrical conductivity
A	−219	−183	poor
B	3550	4830	poor
C	1495	2870	good
D	801	1413	poor when solid, good when molten

6 (a) (i)   What state would you expect substance **D** to be at room temperature?

....

[1]

6 (a) (ii) Suggest what type of bonding substance **D** has.
            Justify your answer using information from the table.

....

....

....

....

[3]

6 (b)   Substance **B** has a giant covalent structure. Explain why it has a high melting point.

....

....

....

[2]

Question 6 continues on the next page

Turn over ▶

6 (c) Substance **A** is oxygen, O_2.
Draw a dot and cross diagram showing the outer electrons in a molecule of oxygen.

[2]

6 (d) Substance **C** is a metal. Explain why it is a good conductor of electricity.

....

....

[2]

[Total 10 marks]

7 (a) Hydrogen can be burned in oxygen and used as a fuel.

$$2H_2 + O_2 \rightarrow 2H_2O$$

7 (a) (i) Calculate the enthalpy change for this reaction.
The bond dissociation energies are given below.

O=O +498 kJ/mol   H-H +436 kJ/mol   O-H +464 kJ/mol

....

....

....

....

Enthalpy change = ... kJ/mol

[4]

7 (a) (ii) Is breaking bonds an exothermic process or an endothermic process?

....

[1]

Question 7 continues on the next page

7 (b) *In this question you will be assessed on the quality of your English, the organisation of your ideas and your use of appropriate specialist vocabulary.*

Evaluate the advantages and disadvantages of using hydrogen fuel cells to power cars instead of petrol.

Remember to finish your answer with a conclusion.

....

....

....

....

....

....

....

....

....

....

....

[6]

[Total 11 marks]

8   Carbon dioxide is a small molecule whose displayed structure is shown below.

$$O=C=O$$

8 (a)   Explain why carbon dioxide has a low boiling point.

....

....

[1]

8 (b)   A student produces carbon dioxide by heating some zinc carbonate ($ZnCO_3$) in a test tube. Zinc oxide (ZnO) is left behind.

8 (b) (i)   State the name for this type of reaction.

....

[1]

8 (b) (ii) Suggest how the student could collect the carbon dioxide produced in this experiment.

....

....

[1]

8 (b) (iii) Describe how the student could show that the gas they had collected was carbon dioxide.

....

....

....

[2]

Question 8 continues on the next page

Turn over ▶

8 (c) The Earth's atmosphere is composed mainly of nitrogen and oxygen, but there is also a small amount of carbon dioxide in air.

8 (c) (i) What percentage of dry air is made up of nitrogen?

....

[1]

8 (c) (ii) Give **one** possible environmental effect of increasing the amount of carbon dioxide in the atmosphere.

....

....

[1]

8 (d) A sample of another compound containing carbon is made up of 96.0 g of calcium, 28.8 g of carbon and 115.2 g of oxygen.

Find the empirical formula of the compound.

....

....

....

....

....

....

Empirical formula = ...

[4]

[Total 11 marks]

9   Metal salts and metal oxides are ionic compounds.
     They can be split up into their elements using electrolysis.

9 (a)   Lead bromide, $PbBr_2$, is a solid at room temperature.
        A student electrolysed a sample of lead bromide using the apparatus shown below.

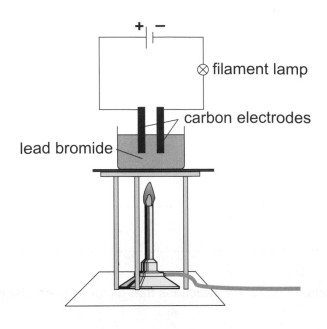

filament lamp

carbon electrodes

lead bromide

9 (a) (i)   Explain why the student only saw the lamp light up after the lead bromide
            had been heated for a while.

            ....

            ....
                                                                                                          [1]

9 (a) (ii) Describe what the student would have observed happening at the positive electrode.
            Explain this observation.

            ....

            ....

            ....

            ....
                                                                                                          [2]

Question 9 continues on the next page

Turn over ▶

9 (b) Aluminium is extracted from aluminium oxide by electrolysis.
The products of this reaction are aluminium metal and oxygen gas.

9 (b) (i) Explain why aluminium cannot be extracted by reduction with carbon.

....

....

[1]

9 (b) (ii) The aluminium oxide is dissolved in molten cryolite for the electrolysis reaction.
Explain why.

....

....

....

[2]

9 (b) (iii) Write the half-equations for the reactions that occur at each electrode.

Negative electrode ....

Positive electrode ....

[4]

9 (b) (iv) The positive electrode is made from carbon. Explain why it has to be regularly replaced.

....

....

....

[1]

[Total 11 marks]

END OF QUESTIONS

Candidate Surname		Candidate Forename(s)	

Centre Number	Candidate Number

Level 1/2 Certificate in Chemistry
Paper 2

Practice Paper
Time allowed: 90 minutes

You must have:
- A ruler.
- A calculator.
- A periodic table.

Total marks:

Instructions to candidates
- Use **black** ink to write your answers.
- Write your name and other details in the spaces provided above.
- Answer **all** questions in the spaces provided.
- In calculations, show clearly how you worked out your answers.

Information for candidates
- The marks available are given in brackets at the end of each question.
- There are 90 marks available for this paper.
- You should answer Question 4 (b) with continuous prose.
  You will be assessed on the quality of your English,
  the organisation of your ideas and your use of appropriate specialist vocabulary.

Advice for candidates
- Read all the questions carefully.
- Write your answers as clearly and neatly as possible.
- Keep in mind how much time you have left.

Turn over ▶

Answer **all** questions

1 A student investigated the properties of the Group 1 metals lithium, sodium and potassium.

1 (a) First the student tested reactivity by putting similar sized pieces of the three metals in water. The student's observations are recorded in the table below.

Metal	Observations
lithium	Fizzes, moves across the surface
sodium	Fizzes strongly, melts into a round ball, moves across the surface
potassium	Fizzes violently, melts into a round ball, moves across the surface, a flame is seen

The student decided that the order of reactivity of the three metals was:

- potassium (most reactive)
- sodium
- lithium (least reactive)

1 (a) (i) Give **two** pieces of evidence from the table that support the student's conclusion.

1. ....

....

2. ....

....

[2]

1 (a) (ii) Explain the pattern of reactivity that the student has noticed.

....

....

....

....

[2]

1 (a) (iii) Complete and balance the chemical equation for the reaction between lithium and water.

.... Li + ... H_2O → .... + ...

[2]

1 (a) (iv) The student added a few drops of universal indicator to the water before adding the lithium.
He observed that the solution became alkaline as the reaction progressed.
Explain why the solution became alkaline.

....

....

[1]

1 (b) The student accidentally mixed up some unlabelled samples of lithium chloride and
potassium chloride. The student decided to do a test to find out which is which,
using a moistened wire loop.

Briefly describe the test that the student could carry out and state what the results would be.

....

....

....

....

[3]

Question 1 continues on the next page

1 (c) For each of the physical properties below, describe briefly how sodium compares to a transition metal like copper:

1 (c) (i) hardness

....

....

[1]

1 (c) (ii) melting point

....

....

[1]

1 (d) The student reads that Group 1 metals will react readily with Group 7 elements. For example, potassium reacts with bromine to form a solid white compound.

1 (d) (i) Name the compound formed in this reaction.

....

[1]

1 (d) (ii) What is the charge on the potassium ion in this compound?

....

[1]

[Total 14 marks]

2   A student wanted to make a chemical cooling pack to use as an alternative to an ice pack for sports injuries. Her idea was to create a strong pack inside which a bag of solid could be burst and dissolved in water.

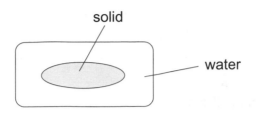

She investigated some possible reactions by mixing 5 g of powdered solid with 50 cm³ of water in a plastic cup and measuring the temperature change.

Chemical used	Temp. at start (°C)	Temp. at end (°C)
ammonium nitrate	22	7
anhydrous copper sulfate	22	31
potassium nitrate	22	15
sodium chloride	22	19

2 (a) (i) One of the chemicals would be unsuitable for use in a sports injury pack. State which chemical and explain why.

....

....

....

[2]

2 (a) (ii) Which chemical would be the best choice for the sports injury pack? Explain your answer.

....

....

....

[2]

Question 2 continues on the next page

Turn over ▶

2 (b) The student insulated all four containers with cotton wool.

2 (b) (i) Explain why the insulation was important in the context of this investigation.

....

....

[2]

2 (b) (ii) Name **two** other things the student did that allowed her to fairly compare the heat change in the reactions.

1. ....

2. ....

[2]

2 (c) (i) Calculate the energy (in J) taken in by each gram of ammonium nitrate as it dissolved. Assume that the solution contains 50 g of water (ignore the mass of solid) and its heat capacity is 4.2 J/g/°C.

The formula for calculating the energy transferred by a reaction is:
Q (J) = m (g) × c × ΔT (°C)

....

....

....

Energy = ... J / g

[2]

2 (c) (ii) Complete the energy level diagram below to show the relative energies of the reactants and products for the ammonium nitrate experiment.

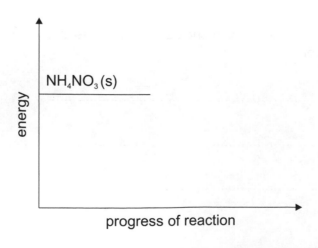

[2]

2 (d) The word equation for the *reversible reaction* between anhydrous copper sulfate and water is shown below.

anhydrous copper sulfate + water ⇌ hydrated copper sulfate

2 (d) (i) Explain what is meant by a *reversible reaction*.

....

....

[1]

2 (d) (ii) What colour would the anhydrous copper sulfate become when it reacted with the water?

Draw a ring around the correct answer.

colourless   red   blue   white

[1]

[Total 14 marks]

Turn over for the next Question

Turn over ▶

3 A student prepared zinc nitrate ($Zn(NO_3)_2$) by dissolving excess zinc oxide (ZnO) in dilute nitric acid (HNO_3). When the reaction had gone to completion, the student filtered the mixture. Then the solution was placed in an evaporating basin and left to crystallise.

3 (a) (i) Write a balanced equation for this reaction, including state symbols.

....

[3]

3 (a) (ii) What name is given to this type of reaction that forms a salt?

....

[1]

3 (b) Explain how the student could tell that the reaction had finished.

....

....

[1]

3 (c) The apparatus that the student used to filter the mixture is shown below. Complete the labels on the diagram.

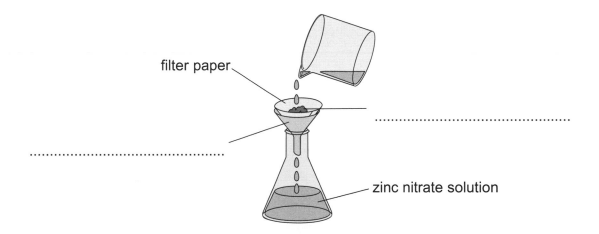

filter paper

....

....

zinc nitrate solution

[2]

3 **(d)** The student used the same method to prepare copper chloride from copper oxide and dilute hydrochloric acid. The equation for this reaction is shown below.

$$CuO(s) + 2HCl(aq) \rightarrow CuCl_2(aq) + H_2O(l)$$

3 **(d) (i)** Explain why copper chloride cannot be prepared from copper and dilute hydrochloric acid.

....

....

[1]

3 **(d) (ii)** Calculate the maximum mass of copper chloride that could be produced if the student used 10.1 g of copper oxide and excess hydrochloric acid.

....

....

....

Answer = ...g

[3]

3 **(e) (i)** The student did the reaction again with a different mass of copper oxide.
He calculated that the reaction should produce 12 g of copper chloride.
The actual mass of copper chloride produced was 8.4 g.

Calculate the percentage yield of this second reaction.

....

....

Answer = ...%

[1]

3 **(e) (ii)** Suggest **one** reason why the yield of the reaction was less than 100%.

....

....

[1]

Question 3 continues on the next page

Turn over ▶

3 (f) Another student performed the same experiment and calculated that their percentage yield of copper chloride was 102%.

3 (f) (i) Suggest **one** reason why the student's yield appeared to be higher than 100%.

....

....

[1]

3 (f) (ii) Suggest how you could check that the reason you suggested in part **(i)** was the cause of the incorrect yield.

....

....

....

[2]

[Total 16 marks]

4   Ammonia can be produced using the Haber process. This process uses the elements
     nitrogen and hydrogen to make ammonia.

4 **(a) (i)**   Write a balanced symbol equation for the Haber process reaction.

                                    $\rightleftharpoons$

....

                                                                                    [2]

4 **(a) (ii)**   The Haber process reaction reaches equilibrium. What does this mean?

....

....

                                                                                    [1]

Question 4 continues on the next page

Turn over ▶

4 (b) *In this question you will be assessed on the quality of your English, the organisation of your ideas and your use of appropriate specialist vocabulary.*

The table below shows the conditions that are used for the Haber process in industry.

Pressure	High (200 atmospheres)
Temperature	450 °C
Catalyst	Iron

Explain why these conditions are used in the Haber process.

....

....

....

....

....

....

....

....

....

....

....

[6]

4 (c) In one part of the Haber process reaction vessel, the reaction mixture is cooled to 30 °C. Explain why the mixture is cooled.

....

....

....

[2]

4 (d) State what happens to any unused nitrogen and hydrogen that leave the reaction vessel.

....

....

[1]

4 (e) A student reads in a textbook that dissolving ammonia in water produces an alkaline solution.

4 (e) (i) Suggest how you could test this and describe what you would observe.

....

....

....

[2]

4 (e) (ii) When this solution reacts with an acid, ammonium salts are formed. Give **one** use of ammonium salts.

....

[1]

[Total 15 marks]

Turn over for the next Question

Turn over ▶

5   A student performed a titration with 0.050 mol/dm³ sulfuric acid in order to determine the concentration of a sample of potassium hydroxide solution.

The apparatus that the student used is shown in the diagram below.

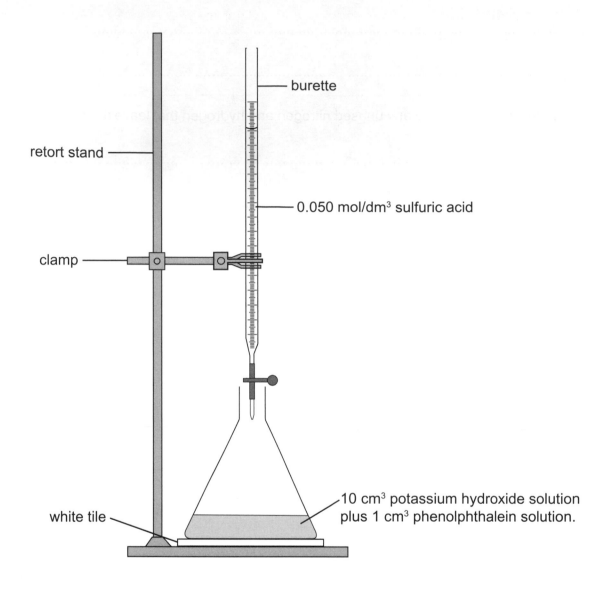

burette

retort stand

0.050 mol/dm³ sulfuric acid

clamp

10 cm³ potassium hydroxide solution plus 1 cm³ phenolphthalein solution.

white tile

5 (a)   Name a piece of apparatus that the student could have used to measure out the potassium hydroxide solution.

....

[1]

5 (b) The student repeated the titration three times. The results are shown in the table below.

5 (b) (i) Calculate the missing values.

Experiment	Initial reading / cm³	Final reading / cm³	Titre Volume / cm³
1	0.20	9.00	8.80
2	3.35	12.25	....
3	1.50	....	8.85

[2]

5 (b) (ii) Why did the student repeat the experiment?

....

....

[1]

Question 5 continues on the next page

122

5 (c) The equation for the reaction between sulfuric acid and potassium hydroxide is shown below.

$$H_2SO_4(aq) + 2KOH(aq) \rightarrow K_2SO_4(aq) + 2H_2O(l)$$

5 (c) (i) A second student performed the same titration. Their average titre volume was 8.80 cm³. How many moles of sulfuric acid reacted with the potassium hydroxide in this titration?

....

....

Number of moles = ...
[2]

5 (c) (ii) How many moles of potassium hydroxide reacted with the acid in this titration?

....

....

Number of moles = ...
[1]

5 (c) (iii) Calculate the concentration of the potassium hydroxide solution in mol/dm³.

....

....

....

Concentration = ... mol/dm³
[2]

5 (c) (iv) The student then performed a similar titration with a solution of sodium hydroxide. He found that the concentration of the sodium hydroxide solution was 0.14 mol/dm³.

Calculate the concentration of this solution in g/dm³.

....

....

Concentration = ... g/dm³
[1]

5 (d) The graph below shows how the pH changed during another titration where potassium hydroxide solution was being added to a weak acid.

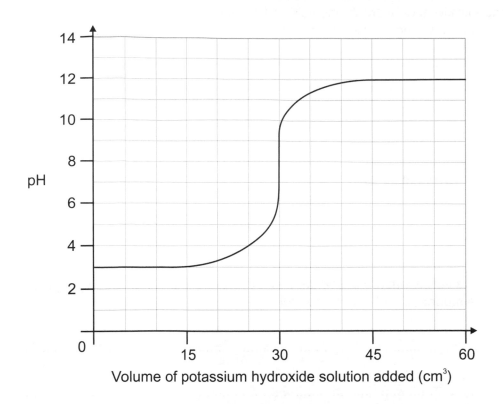

Volume of potassium hydroxide solution added (cm³)

5 (d) (i) What volume of potassium hydroxide solution was needed to neutralise the acid?

....

[1]

5 (d) (ii) Give the pH of the solution when 25.0 cm³ of potassium hydroxide had been added.

....

[1]

5 (d) (iii) State how the titre volume would change if the concentration of the potassium hydroxide solution was doubled. Explain your answer.

You can assume that all the other reactants and conditions are kept the same.

....

....

....

[3]

[Total 15 marks]

Turn over for the next Question

Turn over ▶

6   A student was investigating how the rate of a reaction is affected by concentration.

The student planned to add a measured mass of magnesium ribbon to five different concentrations of hydrochloric acid. She decided to use the apparatus shown below to measure how long it took for each reaction to produce a certain volume of hydrogen gas.

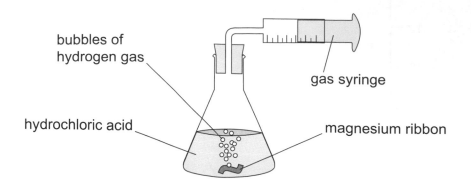

6 (a)   Give **one** safety hazard associated with this experiment and identify a precaution that the student could take to control it.

....

....

[2]

6 (b) (i)   Describe a test that the student could perform to check that the gas produced was hydrogen.

....

....

[1]

6 (b) (ii)   Name the soluble salt present in the flask at the end of the reaction.

....

[1]

6 (c)   The student decided to do a trial run of the experiment to get an idea of how much gas the reactions would produce.

The diagram below shows the volume of gas in the gas syringe at the end of the trial run.

What volume of gas has the student collected?

....

[1]

6 (d) After doing the trial run, the student decided that it would be appropriate to measure how long it took for each reaction to produce 20 cm³ of hydrogen gas.

The results of the experiment are shown in the table below.

Concentration of acid (mol/dm³)	Time taken to produce 20 cm³ of gas (s)
0.2	58
0.4	29
0.6	18
0.8	15
1.0	12

Plot a graph of the student's results on the axes below.
Label the *y* axis and draw a line of best fit.

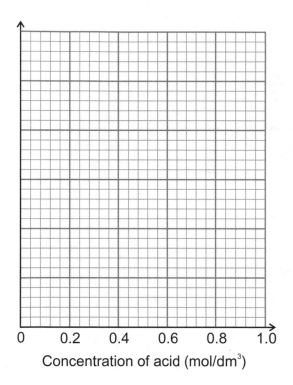

Concentration of acid (mol/dm³)

[4]

6 (e) What conclusions can you draw from the graph?

....

....

....

[2]

Question 6 continues on the next page

Turn over ▶

6 (f) Explain, in terms of collision theory, how concentration affects the rate of a reaction.

....

....

....

....

[2]

6 (g) The student carries out the experiment again using 0.6 mol/dm^3 hydrochloric acid. This time, the student uses the same mass of magnesium powder rather than magnesium ribbon.

The student says "It will take longer than 18 seconds for 20 cm^3 of hydrogen gas to form".

Is the student's prediction correct? Explain your answer.

....

....

....

....

....

[3]

[Total 16 marks]

END OF QUESTIONS

Answers

Section One — Fundamental Ideas in Chemistry

Page 3: States of Matter
1 a)   strong *[1 mark]*, move *[1 mark]*, colder *[1 mark]*.
  b)   E.g. the particles gain energy *[1 mark]*, move around faster
       [1 mark] and go from being close together to being far apart
       [1 mark].
2 a)   D *[1 mark]*
  b)   evaporating *[1 mark]*
  c)   The particles gain energy *[1 mark]* and vibrate more
       [1 mark]. At a certain temperature, the particles have
       enough energy to break free from their positions and the
       solid turns into a liquid/melts. *[1 mark]*.

Page 4: Movement of Particles
1 a)   diffusion *[1 mark]*
  b)   Any two from: e.g. the volume of water / the temperature
       of the water / the size/shape of beaker used
       [1 mark for each correct answer].
2   The particles of ammonia are lighter than particles of
       hydrogen chloride *[1 mark]*, so they move/diffuse through
       the air more quickly *[1 mark]*.
3   Both jars will be the same pale brown colour *[1 mark]*
       because the random motion of the particles means that
       the bromine will eventually diffuse right through the air
       [1 mark].

Page 5: Atoms
1 a)

Particle	Relative mass	Relative Charge
Proton	1	+1
Neutron	1	0
Electron	very small	−1

[3 marks — 1 mark for each correct answer.]

  b) i) Protons are positive and electrons are negative *[1 mark]*.
          The charge on the two protons will balance out the charge
          on the two electrons *[1 mark]*.
     ii)

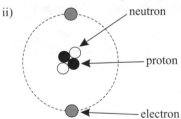

          *[3 marks available — 1 mark for two labelled neutrons in
          the centre, 1 mark for two labelled protons in the centre,
          and 1 mark for two labelled electrons in a shell]*
  c) i)   A substance that is made up of only one type of atom
          [1 mark].
     ii) protons *[1 mark]*

Page 6: Atoms and the Periodic Table
1 a)   column *[1 mark]*,
  b)   E.g. at the right-hand side of the table *[1 mark]*.
2 a)   Na *[1 mark]*
  b) i) bismuth *[1 mark]*
     ii) The two elements will have similar properties *[1 mark]*.
     iii) To the right of the line. Since it doesn't conduct electricity,
          the element must be a non-metal *[1 mark]*.

Page 7: Atomic Mass and Isotopes
1 a)   13 *[1 mark]*
  b) i) The mass number tells you the total number of protons
          and neutrons that the atom contains *[1 mark]*.
     ii) 27 – 13 = **14** neutrons *[1 mark]*
2 a)   Isotopes are atoms of the same element that have a
       different number of neutrons *[1 mark]*.
  b)

Isotope	Mass number	Number of protons	Number of neutrons
^{35}Cl	35	17	18
^{37}Cl	37	17	20

       *[1 mark for correctly stating the mass number and
       number of protons in ^{37}Cl, 1 mark for correctly finding
       the number of neutrons in both ^{35}Cl and ^{37}Cl]*
As ^{35}Cl and ^{37}Cl are isotopes of the same element, they must have the
same number of protons.
  c)   The relative atomic mass is the average mass of the atoms
       of an element *[1 mark]* compared with the mass of one atom
       of carbon-12 *[1 mark]*.

Page 8: Electron Shells
1 a)   2, 2 *[1 mark]*
  b)

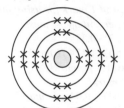

       *[1 mark for the 3rd energy level correct,
       1 mark for the 4th energy level correct.]*
*The important thing here is not exactly where you draw the electrons on
each energy level ring, but that the number on each level is correct —
the shells should be in order as you move away from the nucleus, and you
aren't allowed to put more than two electrons in the first shell, eight in the
second shell, or eight in the third shell.*
  c)   six *[1 mark]*
2 a)   potassium *[1 mark]*
*The atom has (2+8+8+1) = 19 electrons. So you're looking for the element
in the periodic table with atomic number 19 — and that's potassium.*
  b) i) The first energy level only contains one electron *[1 mark]*.
          The first energy level/shell should have two before the next
          energy level/shell is filled *[1 mark]*.
     ii) The second energy level contains nine electrons *[1 mark]*.
          Only eight electrons can fit into the second energy level/shell
          [1 mark].

Page 9: Compounds
1 a)   The two atoms share the electron from the outer shell of
       hydrogen and one of the outer shell electrons of bromine
       [1 mark], forming a covalent bond *[1 mark]*.
  b)   The lithium atom loses the electron in its outer shell,
       becoming a positive ion *[1 mark]*. The bromine atom takes
       this electron into its outer shell, becoming a negative ion
       [1 mark]. The ions are attracted to each other *[1 mark]*.
The attraction of oppositely charged ions is called ionic bonding.
2 a)   To attain a full outer shell/a full outer energy level/the
       electronic structure of the nearest noble gas *[1 mark]*.
  b)   NaCl *[1 mark]*

Page 10: Relative Formula Mass

1 a) The sum of the relative atomic masses of all the atoms that the compound contains *[1 mark]*.

b) M_r of $CaSO_4 = 40 + 32 + (16 \times 4)$
= **136** *[1 mark]*

To find the relative formula mass of a compound, you just write down the atomic masses of all the elements that it contains, multiply each one by the number of atoms of that element that are in the formula, and add it all up.

2 a) M_r of $Z_2CO_3 = 106$
M_r of $CO_3 = 12 + (16 \times 3) = 60$ *[1 mark]*
M_r of $Z_2 = 106 - 60 = 46$
A_r of $Z = 46 \div 2 = $ **23** *[1 mark]*
Element Z is sodium/Na *[1 mark]*

Element Z is sodium because 23 is the A_r of sodium.

b) 106 g *[1 mark]*

The mass of one mole of a substance is equal to its relative formula mass (M_r) in grams.

c) $2.7 \div 106 = $ **0.025** moles *[1 mark]*

3 M_r of $KOH = 39 + 16 + 1 = 56$ *[1 mark]*
Mass of 4 moles of $KOH = 4 \times 56 = 224$ g *[1 mark]*
Extra mass needed $= 224 - 140 = $ **84 g** *[1 mark]*

Page 11: Percentage by Mass and Empirical Formula

1 a) M_r of $NH_4NO_3 = 14 + (1 \times 4) + 14 + (16 \times 3) = 80$
Percentage by mass of N $= ((14 \times 2) \div 80) \times 100 = $ **35%**
[3 marks for correct answer, otherwise 1 mark for correctly calculating the M_r of NH_4NO_3 and 1 mark for using the correct formula for finding percentage by mass]

b) M_r of $Fe_2O_3 = (56 \times 2) + (16 \times 3) = 160$
Percentage by mass of O $= ((16 \times 3) \div 160) \times 100 = $ **30%**
[3 marks for correct answer, otherwise 1 mark for correctly calculating the M_r of Fe_2O_3 and 1 mark for using the correct formula for finding percentage by mass]

2 Division by A_r: Al $= 10.1 \div 27 = 0.374$
Br $= 89.9 \div 80 = 1.124$ *[1 mark]*
Simplest whole number ratio: 1:3 *[1 mark]*
Empirical formula: $AlBr_3$ *[1 mark]*

3 Mass of H $= 1.48 - (0.8 + 0.64) = 0.04$ g *[1 mark]*
Division by A_r: Ca $= 0.8 \div 40 = 0.02$
O $= 0.64 \div 16 = 0.04$
H $= 0.04 \div 1 = 0.04$ *[1 mark]*
Simplest whole number ratio: 1:2:2 *[1 mark]*
Empirical formula: $Ca(OH)_2 / CaO_2H_2$ *[1 mark]*

Pages 12-13: Balancing Equations

1 a) Reactants: methane and oxygen *[1 mark]*
Products: carbon dioxide and water *[1 mark]*

Whenever a question asks you to 'name' a chemical you should really give it's <u>name</u> and not its <u>formula</u>. (If you do give the correct formula, the examiner will usually be kind and give you the mark though.)

b) methane + oxygen → carbon dioxide + water
[1 mark]

c) $CH_4 + 2O_2 \rightarrow CO_2 + 2H_2O$
[1 mark for the correct reactants and products, 1 mark for correctly balancing the equation]

2 2 g of oxygen *[1 mark]*. The mass of the reactants must be equal to the mass of the products / each side of the reaction must have a total mass of 4 g *[1 mark]*.

3 a) i) $2HCl + CuO \rightarrow CuCl_2 + H_2O$ *[1 mark]*
ii) $2HNO_3 + MgO \rightarrow Mg(NO_3)_2 + H_2O$ *[1 mark]*

b) $6HCl + 2Al \rightarrow 2AlCl_3 + 3H_2$
[1 mark for the correct reactants and products, 1 mark for correctly balancing the equation]

4 $Cl_2 + 2KBr \rightarrow Br_2 + 2KCl$
[1 mark for the correct reactants and products, 1 mark for correctly balancing the equation]

5 a) $3CO + Fe_2O_3 \rightarrow 3CO_2 + 2Fe$ *[1 mark]*
b) $2CuO + C \rightarrow 2Cu + CO_2$ *[1 mark]*

6 a) $2Na + Cl_2 \rightarrow 2NaCl$
[1 mark for the correct reactants and products, 1 mark for correctly balancing the equation]

b) $Ca(OH)_2$ (aq) + Na_2CO_3 (s) → $2NaOH$ (aq) + $CaCO_3$ (s)
[1 mark for all state symbols correct]

The word 'precipitate' in the question tells you that the calcium carbonate must be a solid.

c) $2Na$ (s) + $2H_2O$ (l) → $2NaOH$ (aq) + H_2 (g)
[1 mark for correct reactants and products, 1 mark for correctly balancing the equation, 1 mark for all state symbols correct]

Pages 14-15: Calculating Masses in Reactions

1 a) 2Mg ⟶ 2MgO
$2 \times 24 = 48$ $2 \times (24 + 16) = 80$
$48 \div 48 = 1$ g $80 \div 48 = 1.67$ g
$1 \times 10 = 10$ g $1.67 \times 10 = $ **16.7 g**
[3 marks for correct final answer without any working, otherwise 1 mark for correctly calculating both M_rs and 1 mark for dividing through by 48 and multiplying by 10]

b) 4Na ⟶ 2Na₂O
$4 \times 23 = 92$ $2 \times [(2 \times 23) + 16] = 124$
$92 \div 124 = 0.74$ g $124 \div 124 = 1$ g
$0.74 \times 2 = $ **1.48 g** $1 \times 2 = 2$ g
[3 marks for correct final answer, otherwise 1 mark for correctly calculating both M_rs and 1 mark for dividing through by 124 and multiplying by 2]

2 a) Fe_2O_3 ⟶ 2Fe
$(2 \times 56) + (3 \times 16) = 160$ $2 \times 56 = 112$
$160 \div 160 = 1$ g $112 \div 160 = 0.7$ g
$1 \times 20 = 20$ g $0.7 \times 20 = $ **14 g**
[3 marks for correct final answer, otherwise 1 mark for correctly calculating both M_rs and 1 mark for dividing through by 160 and multiplying by 20]

b) 2Al ⟶ Fe_2O_3
$2 \times 27 = 54$ $(2 \times 56) + (3 \times 16) = 160$
$54 \div 160 = 0.3375$ kg $160 \div 160 = 1$ kg
$0.3375 \times 32 = $ **10.8 kg** $1 \times 32 = 32$ kg
[3 marks for correct final answer, otherwise 1 mark for correctly calculating both M_rs and 1 mark for dividing through by 160 and multiplying by 32]

3 Mass of CO produced from 10 g of C at Stage 2:
C ⟶ 2CO
12    $2 \times (12 + 16) = 56$
$12 \div 12 = 1$ g $56 \div 12 = 4.667$ g
$1 \times 10 = 10$ g $4.667 \times 10 = 46.67$ g
Mass of CO_2 made from 46.67 g of CO at stage 3:
3CO ⟶ 3CO₂
$3 \times (12 + 16) = 84$ $3 \times [12 + (16 \times 2)] = 132$
$84 \div 84 = 1$ g $132 \div 84 = 1.571$ g
$1 \times 46.67 = 46.67$ g $1.571 \times 46.67 = $ **73.3 g**
[6 marks for correct final answer. Otherwise: 1 mark for working out the M_rs for Stage 2 and 1 mark for dividing them both by 12. 1 mark for finding correct mass of CO produced in Stage 2. 1 mark for working out the M_rs for Stage 3 and 1 mark for dividing them both by 84]

4 a) 2NaOH ⟶ Na_2SO_4
$2 \times (23 + 16 + 1) = 80$ $(2 \times 23) + 32 + (4 \times 16) = 142$
$80 \div 142 = 0.5634$ g $142 \div 142 = 1$ g
$0.5634 \times 75 = $ **42.3 g** $1 \times 75 = 75$ g
[3 marks for correct final answer, otherwise 1 mark for correctly calculating both M_rs and 1 mark for dividing through by 142 and multiplying by 75]

b) H_2SO_4 ⟶ 2H₂O
$(2 \times 1) + 32 + (4 \times 16) = 98$ $2 \times [(2 \times 1) + 16] = 36$
$98 \div 98 = 1$ g $36 \div 98 = 0.367$ g
$1 \times 50 = 50$ g $0.367 \times 50 = $ **18.4 g**
[3 marks for correct final answer, otherwise 1 mark for correctly calculating both M_rs and 1 mark for dividing through by 98 and multiplying by 50]

Page 16: Percentage Yield and Reversible Reactions

1 a) The yield of a reaction is the amount of product it produces *[1 mark]*.

b) Percentage yield = (1.2 ÷ 2.7) × 100 = **44.4%**
[2 marks for the correct final answer, otherwise 1 mark for correctly substituting the numbers into the percentage yield formula]

2 a) A reaction where the products of the reaction can react to produce the original reactants *[1 mark]*.

b) Some of the methanol that is produced will break down again to give carbon monoxide and hydrogen/the original reactants *[1 mark]*.

3 a) (6 ÷ 15) × 100 = **40%** *[1 mark]*

b) E.g. when the solution was filtered some of the barium sulfate may have been left behind in the beaker. / When the students transferred the solid to a clean piece of filter paper some of it may have been left behind on the first piece of filter paper *[1 mark]*.

Section Two — Bonding and Structure

Pages 17-18: Compounds, Ions and Ionic Bonding

1 A compound is a substance made from atoms of two or more elements *[1 mark]* that are chemically combined together *[1 mark]*.

2 a) magnesium *[1 mark]*

b) fluorine *[1 mark]*

c) Argon *[1 mark]*. Ions are formed when atoms lose or gain electrons to achieve the electronic structure of a noble gas *[1 mark]*. Argon is a noble gas. / Argon already has the electronic structure of a noble gas. / Argon already has a full outer electron shell *[1 mark]*.

3 a) Non-metals form negative ions/form ions by gaining electrons *[1 mark]*.

b) −2 *[1 mark]*

Elements in Group 6 have six electrons in their outer shells, so they need to gain two more electrons to fill up their outer shells. This means that they will form −2 ions.

4 One electron moves/is transferred *[1 mark]* from the outer shell of the lithium atom *[1 mark]* to the outer shell of the chlorine atom *[1 mark]*.

5 a) a lattice *[1 mark]*

b) The lattice is held together by strong electrostatic forces of attraction *[1 mark]* between oppositely charged ions *[1 mark]*. These forces act in all directions *[1 mark]*.

c)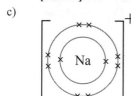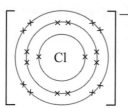

[1 mark for the sodium ion having the correct electron arrangement, 1 mark for the chloride ion having the correct electron arrangement, 1 mark for both charges correct]

6 a) When the ionic compound is solid the ions are not free to move *[1 mark]*. When dissolved in water the ions are free to move in the solution *[1 mark]*. The ions must be free to move in order to carry an electric current *[1 mark]*.

b) A large amount of energy is needed *[1 mark]* to overcome the strong forces of attraction between the oppositely charged ions in ionic compounds *[1 mark]*.

Pages 19-20: Covalent Bonding

1 a) A covalent bond is a shared pair of electrons *[1 mark]*. Atoms form covalent bonds to achieve a full outer shell of electrons / the electronic structure of a noble gas *[1 mark]*.

b) C *[1 mark]*

2 a) i)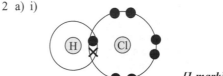

[1 mark]

ii)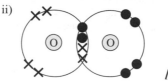

[1 mark]

iii)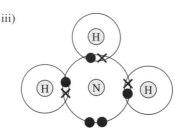

[1 mark]

b) A pair of electrons is shared between the two atoms *[1 mark]*. One electron is from the hydrogen atom and one electron is from the chlorine atom *[1 mark]*.

3 a) i) E.g.
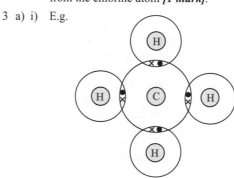

[1 mark for the correct arrangement of atoms, 1 mark for showing the four shared electron pairs]

ii) 4 *[1 mark]*

b) 3 *[1 mark]*

c) i) oxygen *[1 mark]*

ii) non-metal *[1 mark]*

4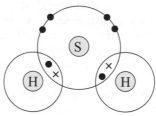

[2 marks available — 1 mark for both shared pairs, 1 mark for the non-bonding electrons.]

Oxygen and sulfur are both in Group 6, and both have six electrons in their outer shells. So all you need to do to draw H_2S is to think about how you would draw H_2O and replace the oxygen atoms with sulfur.

Pages 21-22: Covalent Substances: Two Kinds

1 a) D *[1 mark]* — silicon dioxide is a giant covalent structure *[1 mark]* so will have a high melting point and won't conduct electricity *[1 mark]*.

b) A *[1 mark]* — chlorine is a simple molecular substance *[1 mark]* so will have a low melting point (and won't conduct electricity) *[1 mark]*.

2 a) Simple molecular substances have weak intermolecular forces between their molecules *[1 mark]*. Only a little energy is needed to overcome these forces and separate the molecules *[1 mark]*.

b) The molecules are not charged / do not contain any ions/ delocalised electrons to carry the charge *[1 mark]*.

3 a) How to grade your answer:
0 marks: No relevant points are made.
1-2 marks: A brief attempt is made to explain one or two of the properties from the table in terms of structure and bonding.
3-4 marks: Some explanation of three or four of the properties from the table is given in terms of structure and bonding. The answer has a logical structure and spelling, grammar and punctuation are mostly correct.
5-6 marks: A clear and detailed explanation of five or all of the properties from the table is given in terms of structure and bonding. The answer has a logical structure and uses correct spelling, grammar and punctuation.

Here are some points your answer may include:

Diamond:
Each carbon atom in diamond forms four covalent bonds in a rigid giant structure. This makes diamond very hard.
All of the atoms in diamond are bonded to each other by strong covalent bonds. These take a lot of energy to break, so diamond has a very high melting point.
Each carbon atom in diamond forms four covalent bonds, so there are no free electrons. This is why diamond doesn't conduct electricity.

Graphite:
Each carbon atom in graphite forms three covalent bonds, creating layers of carbon atoms that can slide over each other. The carbon layers are only held together weakly, which is what makes graphite soft and slippery.
The atoms in graphite are bonded to each other by strong covalent bonds within the layers. These take a lot of energy to break, so graphite has a very high melting point.
Only three out of each carbon atom's four outer electrons are used in bonds, so graphite has lots of delocalised electrons, which means that it can conduct electricity.

b) C *[1 mark]*

Page 23: Fullerenes and Nanoscience

1 a) carbon *[1 mark]*, hexagon *[1 mark]*
b) i) E.g. to reinforce graphite in tennis rackets / to make stronger, lighter building materials / making electronic circuits for tiny computer chips *[1 mark]*.
ii) Any two from: e.g. to deliver drugs into the body / in lubricants / to reinforce graphite in tennis rackets / to make stronger, lighter building materials / making electronic circuits for tiny computer chips *[2 marks — 1 mark for each correct answer not already given in part b)i)]*

2 a) i) E.g. the zinc oxide nanoparticles will absorb UV rays *[1 mark]* but the new sun cream would not leave white marks on the skin *[1 mark]*.
ii) 30 nm *[1 mark]*.
b) Nanoparticles have a much higher surface area to volume ratio than the same material in bulk *[1 mark]*.

Section Three — Air and Water

Page 24: Air

1 a) nitrogen *[1 mark]*
b) 21% *[1 mark]*
c) Any two from: carbon dioxide / water vapour / argon / neon *[1 mark for each correct answer]*.
2 a) fractional distillation *[1 mark]*
b) Their boiling points *[1 mark]*.
c) Test the gas with a glowing spill *[1 mark]*.
Oxygen will relight the spill *[1 mark]*.

Page 25: Oxygen and Burning

1 a)

Element	Flame colour when burnt	Oxide formed	Acid-base character of oxide
sodium	Yellow	Na_2O	Alkaline
magnesium	White	MgO	Alkaline
carbon	Orange/yellow	CO_2	Acidic
sulfur	Blue	SO_2	Acidic

[1 mark for each correct answer]
b) Being able to react as either an acid or a base *[1 mark]*.
2 a) iron + oxygen $\rightarrow$ iron oxide *[1 mark]*
b) Oxidised, because oxidation is the addition of oxygen *[1 mark]* and in this reaction oxygen is added to the iron *[1 mark]*.
3 Alkaline *[1 mark]*, because when metal oxides dissolve in water they give alkaline solutions *[1 mark]*.

Page 26: Air Pollution

1 a) Carbon monoxide can form if the fuel is burnt without enough oxygen *[1 mark]*.
b) Carbon monoxide combines with haemoglobin in blood cells *[1 mark]* meaning that the blood can carry less oxygen *[1 mark]*.
2 a) E.g. in a car engine *[1 mark]*.
b) $3NO_2 + H_2O \rightarrow 2HNO_3 + NO$
[1 mark for each correct number]
3 a) Most fossil fuels contain sulfur impurities *[1 mark]*. When the fuel burns, the sulfur reacts with oxygen in the air to form sulfur dioxide *[1 mark]*.
b) E.g. sulfur dioxide causes acid rain/reacts with moisture in clouds to form sulfuric acid, which falls as acid rain *[1 mark]*. Acid rain could cause the lake to become acidic *[1 mark]*, killing fish/insects/plants/wildlife *[1 mark]*.

Pages 27-28: Water Quality

1 a) E.g. dip blue cobalt chloride paper into the sample *[1 mark]*. It will turn pink if the sample contains water *[1 mark]*.
b) 100 °C *[1 mark]*.
2 a) i) Treating (sea) water to remove the salt *[1 mark]*.
ii) E.g. countries with low rainfall might not have much freshwater/river water/groundwater *[1 mark]* so they might need to make drinking water from sea water *[1 mark]*.
b) E.g. distillation *[1 mark]*.
3 The water is filtered/screened/passed through filter beds *[1 mark]* to remove any solids *[1 mark]*. Chlorine is then added to the water *[1 mark]* to sterilise it / kill microbes *[1 mark]*.
4 How to grade your answer:
0 marks: No advantages and disadvantages given.
1-2 marks: Brief description of at least one advantage and one disadvantage, with no attempt at a conclusion.
3-4 marks: Some description of at least two advantages and two disadvantages are given, and some attempt is made to give a conclusion. The answer has a logical structure and spelling, grammar and punctuation are mostly correct.
5-6 marks: A clear description of at least two advantages and disadvantages is given and a sensible conclusion is made. There is a logical structure and the answer uses correct spelling, grammar and punctuation.

Here are some points your answer may include:

Advantages:

Fluoride can help to reduce tooth decay.

Chlorine can kill microbes and therefore prevent disease.

Disadvantages:

Some studies have linked adding chlorine to water with an increase in certain cancers. / Chlorine can react with other natural substances in water to produce toxic by-products which some people think could cause cancer.

In high doses, fluoride can cause cancer and bone problems. There is concern over whether it's right to mass medicate, since people can't choose whether their tap water has fluoride added to it.

Conclusion:

E.g. overall, adding chlorine and fluoride is good because they have several health benefits. / Overall, adding chlorine and fluoride isn't right, because they may cause health problems, and people should have the right to decide which chemicals are put into the water they drink.

5 a) Any two from: e.g. to improve the taste of the water / to improve the quality of the water / to remove bacteria from the water *[1 mark for each correct answer]*.

b) Hard water contains lots of calcium/magnesium ions *[1 mark]*. The ion exchange resin contains lots of sodium/ hydrogen ions *[1 mark]*. These are exchanged for the calcium/magnesium ions in the water *[1 mark]*.

Page 29: Rust

1 a) corrosion *[1 mark]*, water *[1 mark]*, air *[1 mark]*

b) tube B *[1 mark]*

c) The painted nail would not rust *[1 mark]*, because the paint creates a barrier that keeps out water and oxygen *[1 mark]*.

d) Zinc is more reactive than iron *[1 mark]*, so the zinc will be oxidised instead of the iron *[1 mark]*.

e) e.g. greasing *[1 mark]*

Section Four — The Periodic Table and Metals

Page 30: More About The Periodic Table

1 Chlorine has one more electron shell than fluorine *[1 mark]*. This means that its outer electron shell is further from the nucleus than fluorine's *[1 mark]* and more shielded by inner electrons *[1 mark]*. So it is more difficult for the nucleus of the chlorine atom to attract electrons / for the chlorine atom to gain electrons *[1 mark]*.

2 a) In order of increasing atomic (proton) number *[1 mark]*.

b) Cl and Br *[1 mark]*

If you put a tick in more than one box, you won't get the mark.

c) Sodium and potassium are in the same group, so they have the same number of electrons in their outer shell *[1 mark]*.

d) They have a full outer shell of electrons *[1 mark]* so they are very stable/don't need to lose or gain electrons in a reaction to obtain a full outer shell *[1 mark]*.

Page 31: Group 1 — The Alkali Metals

1 a) Lithium is less dense than water *[1 mark]*.

b) alkaline *[1 mark]*

c) $2Li (s) + 2H_2O (l) \rightarrow 2LiOH (aq) + H_2 (g)$
[1 mark for H_2, 1 mark for correct balancing]

2 As the atomic number of the alkali metals increases, the outer electron is further from the nucleus *[1 mark]* and is more shielded by the inner electrons *[1 mark]*. This means that the outer electron is less strongly attracted to the nucleus *[1 mark]*, so it is more easily lost *[1 mark]*.

3 a) Sodium iodide is a white solid *[1 mark]*. The charge on the sodium ion in sodium iodide is +1 *[1 mark]*.

b) It would dissolve *[1 mark]* to give a colourless solution *[1 mark]*.

Page 32: Group 7 — The Halogens

1 a) The compound is iron bromide *[1 mark]*. The type of bonding present in iron bromide is ionic bonding *[1 mark]*.

b) −1 *[1 mark]*

2 a) Fluorine *[1 mark]*. Fluorine is above chlorine in Group 7 *[1 mark]*, and the further up Group 7 an element is, the lower its boiling point *[1 mark]*.

b) iodine / astatine *[1 mark]*

3 a) $Br_2 + 2KI \rightarrow I_2 + 2KBr$
[1 mark for the correct reactants and products, 1 mark for correct balancing]

b) E.g. bromine is more reactive than iodine *[1 mark]*, so it will displace iodine from the potassium iodide solution *[1 mark]*.

Page 33: Transition Elements

1 a) Metal B. Transition elements are harder than Group 1 elements *[1 mark]*.

b) Metal D. Transition elements have higher melting points than Group 1 elements *[1 mark]*.

c) Metal E. Transition elements have higher densities than Group 1 elements *[1 mark]* and are much less reactive than Group 1 elements *[1 mark]*.

2 a) The iron is acting as a catalyst *[1 mark]*.

b) i) E.g. iron can form ions with different charges *[1 mark]*.

ii) It is brightly coloured *[1 mark]*.

Pages 34-35: Reactions of Metals and the Reactivity Series

1 a) i) magnesium sulfate *[1 mark]*, hydrogen *[1 mark]*

ii) Apply a lighted splint to a sample of the gas *[1 mark]*. If the gas burns with a squeaky pop, it must be hydrogen *[1 mark]*.

b) least vigorous   B → A → C → D   most vigorous
[1 mark for any two letters correct, 2 marks for all four letters correct]

The clue is in the diagrams — the reaction that produces the most gas will push the gas syringe out the furthest.

c) zinc *[1 mark]*

d) sodium + water → sodium hydroxide + hydrogen
[1 mark for sodium hydroxide, 1 mark for hydrogen]

2 a) iron / tin / lead *[1 mark]*

The unidentified metal must be more reactive than copper (as it displaces copper from copper sulfate), but less reactive than zinc (as it doesn't displace zinc from zinc sulfate).

b) The unidentified metal displaced copper from its salt *[1 mark]*

c) $Cu^{2+} (aq) + Zn (s) \rightarrow Zn^{2+} (aq) + Cu (s)$
[1 mark]

3 a) There would be no reaction *[1 mark]*.

b) i) hydrogen *[1 mark]*

ii) Because iron is more reactive than hydrogen *[1 mark]*, but copper is less reactive than hydrogen *[1 mark]*.

Pages 36-38: Getting Metals From Rocks

1 a) Iron is lower in the reactivity series than carbon *[1 mark]*, so carbon can take oxygen away from/reduce iron *[1 mark]*.

b) reduction *[1 mark]*

c) iron(III) oxide + carbon → iron + carbon dioxide *[1 mark]*

d) Gold is unreactive *[1 mark]* so it doesn't form compounds/ is found in the Earth as the pure metal *[1 mark]*.

2 a) i) electrolysis *[1 mark]*

ii) Electrolysis is expensive *[1 mark]* because it uses a lot of energy *[1 mark]*.

b) Carbon cannot be used to extract/take oxygen away from/ reduce any metals that are more reactive than itself *[1 mark]*.

3 a) smelting *[1 mark]*

b) It is impure *[1 mark]*, so it isn't a good conductor of electricity *[1 mark]*.

4 a)   copper sulfate solution — D
        copper ions — C
        pure copper electrode — A
        impure copper electrode — B
        [1 mark for two correct answers, 2 marks for all four correct]

b) i) $Cu \rightarrow Cu^{2+} + 2e^-$
        [1 mark]

   ii) $Cu^{2+} + 2e^- \rightarrow Cu$
        [1 mark]

c)   Electrode X is the anode *[1 mark]*. It decreased in mass, and the mass of the anode decreases as copper ions are lost from it and move towards the cathode *[1 mark]*.

5 a)   Because iron is more reactive than copper *[1 mark]*.

b) $Fe + Cu^{2+} \rightarrow Cu + Fe^{2+}$
     [1 mark for correct reactants, 1 mark for correct products].

c)   iron *[1 mark]*

6 a)   Bioleaching uses bacteria to separate copper from copper sulfide *[1 mark]*. The leachate (the solution produced by the bacteria) contains copper (ions) which can then be extracted *[1 mark]*.

b)   phytomining *[1 mark]*

c)   The supply of copper-rich ores is limited *[1 mark]*. These alternative methods can extract copper from low-grade ores / from the waste that is currently produced when copper is extracted *[1 mark]*.

Page 39: Impacts of Extracting Metals

1 a)   How to grade your answer:

     0 marks: No advantages and disadvantages given.

     1-2 marks: Brief description of at least one advantage and one disadvantage, with no attempt at a conclusion.

     3-4 marks: At least two advantages and two disadvantages are given, and some attempt is made to give a conclusion. The answer has a logical structure and spelling, grammar and punctuation are mostly correct.

     5-6 marks: The answer gives at least two advantages and disadvantages and finishes with a sensible conclusion. There is a logical structure and the answer uses correct spelling, grammar and punctuation.

     Here are some points your answer may include:

     Advantages:
     Mining metal ores allows useful products to be made from the metal.
     The mines provide jobs for workers.
     Mining brings money into the local area/brings improved local services.
     Disadvantages:
     Mining can be very noisy.
     Mines scar the landscape.
     Mining can lead to a loss of wildlife habitats.
     Abandoned mine shafts can be dangerous.
     Conclusion:
     E.g. overall, the benefits of mining outweigh the problems, because the metals obtained are essential for making many products. / Overall, mining is a bad way to obtain metals because it is far more damaging to the environment than other ways of obtaining metals.

b)   Any three from: e.g. recycling metals uses much less energy than mining and extracting/saves energy / conserves finite resources / reduces the amount of waste sent to landfill / saves fossil fuels/reduces pollution caused by burning fossil fuels during mining/extraction / saves money
     [1 mark for each correct answer].

Pages 40-41 Metals

1 a)   pattern *[1 mark]*, delocalised *[1 mark]*

b)   E.g.

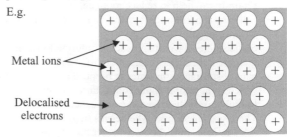

Metal ions

Delocalised electrons

     [1 mark for regular structure of atoms, 1 mark for delocalised electrons]

c) i)   Metals conduct electricity because the (outer) electrons of their atoms are delocalised, *[1 mark]* so they are free to move throughout the entire structure *[1 mark]*.

   ii)   The layers of atoms are able to slide over each other *[1 mark]*. This means the metals can be bent *[1 mark]*.

2 a)   transition metals/elements *[1 mark]*

b)   E.g. they are malleable/can be bent into shape *[1 mark]* and they are good electrical conductors *[1 mark]*.

3 a)   Any two from: e.g. copper does not react with water / copper can be bent and shaped / copper is very hard *[1 mark for each correct answer]*.

b)   impurities *[1 mark]*

4 a)   A and B *[1 mark]*

b)   A *[1 mark]*, because it took the least time for the end that wasn't near the heat source to heat up *[1 mark]*.

Page 42: Alloys

1 a)   An alloy is a mixture of two or more metals (or a metal and a non-metal) *[1 mark]*.

b)   In pure metals like copper the regular layers of atoms are able to slide over each other *[1 mark]*. This means copper bends easily *[1 mark]*. However, in alloys such as brass there are atoms of more than one size *[1 mark]*. This distorts the layers and prevents them from being able to slide over each other *[1 mark]*.

2 a)   low/small *[1 mark]*

b)   high *[1 mark]*

c)   stainless *[1 mark]*

3 a)   It can be bent/shaped when cool *[1 mark]*, but it returns to its original shape when heated *[1 mark]*.

b)   The warmth of the mouth causes the nitinol to try to return to its original shape *[1 mark]*. As it changes shape, it pulls the teeth with it *[1 mark]*.

Section Five — Acids, Bases and Reaction Rates

Page 43: Acids and Bases

1 a)   0 to 14 *[1 mark]*

b)   neutral *[1 mark]*

c)   strong alkali *[1 mark]*

d)   orange *[1 mark]*

2 a) H^+ ions/hydrogen ions *[1 mark]*

b) i)   neutralisation *[1 mark]*

   ii) H^+ (aq) $+ OH^-$ (aq) $\rightarrow H_2O$ (l)
     [1 mark for all reactants and products correct, 1 mark for state symbols]

c)   a salt *[1 mark]*

d)   The indicator will turn green (because universal indicator is green in neutral solutions) *[1 mark]*.

When all of the acid has reacted with the alkali the solution will be neutral — so the universal indicator will go green.

Page 44: Oxides, Hydroxides and Ammonia

1 a) magnesium sulfate + water *[1 mark]*
  b) aluminium chloride + water *[1 mark]*
2 alkaline *[1 mark]*, fertilisers *[1 mark]*
3 a) nitric acid + copper oxide → copper nitrate + water *[1 mark]*
  b) $2HCl + ZnO \rightarrow ZnCl_2 + H_2O$
     [1 mark for all reactants and products correct, 1 mark for correct balancing]
  c) sulfuric acid *[1 mark]*
4 a) lithium hydroxide *[1 mark]*
  b) iron hydroxide, iron oxide, lithium hydroxide
     [1 mark for all three correct]

All metal oxides and hydroxides are bases. Metal hydroxides that will dissolve in water are alkalis.

Page 45: Titrations

1 a) E.g. the student is filling the burette above eye level, so the acid could splash into her eyes *[1 mark]*. The student should lower the equipment to below eye level to fill the burette/use a funnel to fill the burette/wear safety goggles *[1 mark]*.
Any other sensible hazard that you've spotted in the diagram would also be fine here. To get both marks, your suggestion for minimising the risk must match the hazard you've written about.
  b) Universal indicator changes colour gradually *[1 mark]* — titrations need a definite colour change when the solution changes between acid and alkali *[1 mark]*.
  c) So that she did not miss the end-point/colour change. / So that she could be sure she had accurately recorded the exact volume of acid added at the end-point/colour change *[1 mark]*.
  d) i) The student should read the level of the liquid in the burette against the scale at eye level *[1 mark]* from the bottom of the meniscus *[1 mark]*.
    ii) E.g. repeating the titration and taking a mean improves the reliability of the final result / helps to reduce the effect of random errors on the result *[1 mark]*.

Page 46: Titration Calculations

1 a) $0.100 \times (20.0 \div 1000)$
     = **0.002 moles**
     [2 marks for correct answer, otherwise 1 mark for multiplying concentration by volume in cm³]
  b) $0.002 \div 2 =$ **0.001 moles** *[1 mark]*
  c) $0.001 \div (50.0 \div 1000)$
     = **0.02 mol/dm³**
     [2 marks for correct answer, otherwise 1 mark for dividing number of moles by volume in cm³]
  d) M_r of $Ca(OH)_2$ is $(1 \times 40) + (2 \times 16) + (2 \times 1) = 74$
     Concentration = $0.02 \times 74 =$ **1.48 g/dm³**
     [2 marks for correct answer, otherwise 1 mark for calculating the M_r of $Ca(OH)_2$]
If you got parts a), b) or c) wrong here, you still get 1 mark in each of the following parts of the question if you've used the correct method.
2 a) Moles KOH = $0.100 \times (30.0 \div 1000) = 0.003$
     Reaction equation shows that 2 moles of KOH reacts with 1 mole of H_2SO_4, so 0.003 moles of KOH reacts with $0.003 \div 2 = 0.0015$ moles of H_2SO_4
     Concentration = $0.0015 \div (10.0 \div 1000)$
     = **0.15 mol/dm³**
     [5 marks for correct answer. Otherwise: 1 mark for multiplying the concentration of the KOH solution by its volume in cm³, 1 mark for saying that there are 0.003 moles of KOH, 1 mark for saying that there are 0.0015 moles of H_2SO_4 and 1 mark for dividing the number of moles of H_2SO_4 by its volume in cm³]
The first part of your calculation tells you that there are 0.0015 moles of H_2SO_4 in 10.0 cm³ of the solution. Then you can use this to work out how many moles of H_2SO_4 there would be in 1 dm³ (1000 cm³) of solution — this is the concentration in mol/dm³.

  b) M_r of H_2SO_4 is $(2 \times 1) + (1 \times 32) + (4 \times 16) = 98$
     Concentration = $0.15 \times 98 =$ **14.7 g/dm³**
     [2 marks for correct answer, otherwise 1 mark for calculating the M_r of H_2SO_4]

Pages 47-48: Making Salts

1 a) i) B *[1 mark]*
    ii) C *[1 mark]*
    iii) A *[1 mark]*
    iv) D *[1 mark]*
  b) This method is not safe, because sodium metal is too reactive/reacts explosively with acids *[1 mark]*.
2 a) silver nitrate + sodium chloride
     → silver chloride + sodium nitrate *[1 mark]*
  b) The silver chloride must be filtered out of the solution *[1 mark]*. It needs to be washed and then dried on filter paper *[1 mark]*.
  c) E.g. the removal of harmful ions from drinking water. / The removal of calcium and magnesium ions from hard water / Treating water for drinking *[1 mark]*.
3 a) E.g. the zinc oxide would stop dissolving in the acid. / The excess zinc oxide would sink to the bottom of the flask and stay there *[1 mark]*.
  b) zinc oxide + nitric acid → zinc nitrate + water *[1 mark]*
  c) Evaporate some of the water (to make the solution more concentrated) *[1 mark]* and then leave the rest to evaporate very slowly *[1 mark]*.
  d) i) Sodium hydroxide is soluble, so you can't tell/see when the reaction is finished *[1 mark]*. This means you can't add an excess of the base to the acid and filter out what's left *[1 mark]*.
    ii) E.g. add the sodium hydroxide to a known volume of the nitric acid gradually/titrate a known volume of the nitric acid with the sodium hydroxide *[1 mark]* using an indicator to show when the reaction has finished/the acid has been neutralised *[1 mark]*. Then mix exactly the same volumes of the acid and the alkali, but without the indicator, to produce a pure solution *[1 mark]*.

Pages 49-50: Metal Carbonates and Limestone

1 a) calcium carbonate *[1 mark]*
  b) by quarrying *[1 mark]*
  c) i) calcium oxide *[1 mark]*
    ii) carbon dioxide *[1 mark]*
    iii) thermal decomposition *[1 mark]*
  d) water *[1 mark]*, carbon dioxide *[1 mark]*
  e) i) Calcium carbonate/limestone is a base *[1 mark]*, so farmers can spread it on fields to neutralise soils that are acidic *[1 mark]*.
    ii) Any two from: e.g. as a building material / making cement / making glass / making iron / making lime
       [1 mark for each use]
2 a) magnesium nitrate *[1 mark]*
  b) Bubble the gas through limewater/calcium hydroxide solution *[1 mark]*. The solution will turn cloudy if there is carbon dioxide in the gas *[1 mark]*.
3 a) i) $CuCO_3 \rightarrow CuO + CO_2$
       [1 mark for reactants correct, 1 mark for products correct]
This is yet another thermal decomposition reaction.
    ii) copper oxide *[1 mark]*
  b) i) limewater *[1 mark]*
    ii) $Ca(OH)_2 + CO_2 \rightarrow CaCO_3 + H_2O$
       [1 mark for reactants correct, 1 mark for products correct]
  c) The sodium carbonate did not thermally decompose *[1 mark]* because the Bunsen burner did not reach a high enough temperature for the reaction to happen *[1 mark]*.

Pages 51-52 Rates of Reaction

1 a) Z *[1 mark]*, because curve Z has the steepest initial gradient/ flattens off earliest *[1 mark]*.

b) The same mass of marble was used in each experiment (and the acid was in excess) *[1 mark]*.

2 a) The mass of the reaction mixture will decrease *[1 mark]*.
One of the products of this reaction is carbon dioxide. As the reaction progresses, the carbon dioxide is released into the air, so the mass of the reaction mixture will decrease.

b) Change in mass between 10 and 25 seconds = 4 − 2 = 2 g
Rate = amount of reactant used ÷ time
= 2 g ÷ 15 s
= **0.13 g/s**
[3 marks for correct answer, otherwise 1 mark for finding that 2 g of reactant was used between 10 s and 25 s and 1 mark for dividing the amount of reactant used by time]

3 a) 50 in the third column of the table (the result of Experiment 2 with 0.5 mol/dm³ HCl) should be circled *[1 mark]*.

b) i)

Mean
94
64
45.5
20
9

[1 mark for each correct mean]
To find the mean for any concentration, you just add the results from Experiment 1 and Experiment 2 and divide by two.

ii)
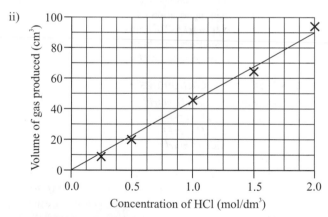

[4 marks available — 1 mark for labelling the x axis, 2 marks for plotting all the data points correctly or 1 mark for plotting any four of the data points correctly, 1 mark for a suitable line of best fit]

c) 2 mol/dm³ *[1 mark]*, because the largest mean volume of gas (94 cm³) was produced in the first minute of that reaction *[1 mark]*.

d) i) Gas syringe *[1 mark]*
ii) stopwatch / timer / balance / measuring cylinder / pipette *[1 mark]*

Pages 53-56: Collision Theory and Catalysts

1 a) Increasing the pressure means there are more particles/ molecules/atoms of gas in a given volume/space / the particles are closer together *[1 mark]*, so collisions happen more frequently *[1 mark]*.

b) i) activation energy *[1 mark]*
ii) By increasing the temperature *[1 mark]*.

2 a) E.g.
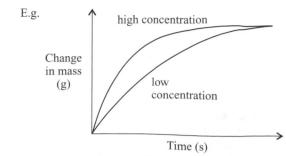

[1 mark for each correct line]
The high concentration curve should go up more steeply and flatten out earlier than the low concentration curve, but both curves should head towards the same end point.

b) E.g. acid spitting/splashing during the reaction *[1 mark]* — wear safety goggles/gloves to protect the eyes/hands *[1 mark]*.
Any sensible hazard and matching precaution will get you the marks here.

3 a) i) The rate of reaction increases *[1 mark]*.
ii) When the temperature is increased the particles move faster *[1 mark]*. This means that they will collide more frequently *[1 mark]*. Reactions will only happen if the particles collide with enough energy *[1 mark]*. Faster movement means more of the particles will collide with enough energy to react *[1 mark]*.

b) E.g. the experiment could be repeated at least three times *[1 mark]*.

c) i) Increasing the concentration of hydrochloric acid/ HCl increases the rate of this reaction. / Decreasing the concentration of hydrochloric acid/HCl decreases the rate of this reaction *[1 mark]*.
ii) When the solution is more concentrated, there are more particles of reactant in the same volume of solution *[1 mark]*, so collisions are more frequent *[1 mark]*.

4 a) A, because powdered marble has a larger surface area than large marble chips *[1 mark]*, so collisions between the surface and the particles in solution are more frequent *[1 mark]*.

b) E.g.
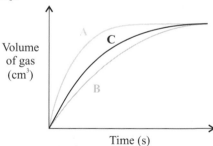

[1 mark for a curve labelled C drawn between A and B]

c) i) E.g. the concentration of the acid / the temperature of the acid *[1 mark for each correct variable]*.
ii) E.g. the amount of gas produced would halve *[1 mark]*.

d) E.g. measure the loss in mass of the reaction (using a mass balance) *[1 mark]*.
Any sensible suggestion for how to measure the rate of this reaction would be fine here.

5 a) A catalyst is a substance which speeds up a reaction *[1 mark]* without being changed or used up in the reaction. *[1 mark]*.

b) A — this reaction has the steepest graph/flattens out soonest *[1 mark]*, so it shows the faster of the two reactions (which is the one with the catalyst added) *[1 mark]*.

6 a) E.g. Catalysts speed up reactions *[1 mark]*, so a chemical plant won't have to operate for as long to produce the same amount of product *[1 mark]*. Also catalysts allow reactions to take place at lower temperatures, saving energy *[1 mark]*.

b) E.g. catalysts can be expensive to buy. / Catalysts often need to be removed from the product and cleaned. / Different reactions need different catalysts, so a plant making more than one product will need more than one catalyst *[1 mark]*.

Section Six — Crude Oil and Organic Chemistry

Page 57: Crude Oil

1 a) A mixture is a substance that consists of two or more elements or compounds that aren't chemically bonded to each other *[1 mark]*.

b) hydrocarbons *[1 mark]*

c) G *[1 mark]*

d) How to grade your answer:

0 marks: No relevant points are made.

1-2 marks: There is a brief outline of the process of fractional distillation. The answer contains little or no explanation of how the process works.

3-4 marks: There is some description of the process of fractional distillation. The answer contains some explanation of how the process works. The answer has a logical structure and spelling, grammar and punctuation are mostly correct.

5-6 marks: There is a clear and detailed description of the process of fractional distillation. The answer contains a full explanation of how the process works. The answer has a logical structure and uses correct spelling, grammar and punctuation.

Here are some points your answer may include:

Crude oil is heated until it evaporates.

The vapours enter the fractionating column and rises up it.

The column has a temperature gradient. The temperature decreases the further up the column you go.

Different fractions have different boiling points.

Fractions made up of large (hydrocarbon) molecules have high boiling points, so they condense and drain out near the bottom of the column.

Fractions made up of small (hydrocarbon) molecules have low boiling points, so they condense near the top of the column, where it's cooler.

The various fractions are constantly tapped off at the different levels where they condense.

Each fraction contains molecules with a similar number of carbon atoms to each other.

Page 58: Properties and Uses of Crude Oil

1 a) i)

H H H
| | |
H– C– C– C– H
| | |
H H H   *[1 mark]*

propane *[1 mark]*

ii) Alkane 1. The shorter the hydrocarbon molecule, the lower the boiling point *[1 mark]*.

b) i) C_nH_{2n+2} *[1 mark]*

ii) homologous series *[1 mark]*

c) They are saturated — all their carbon-carbon bonds are single covalent bonds *[1 mark]*.

2 Any two from: e.g. they are highly flammable and could catch fire in the hot engine. / They would not be viscous enough to lubricate the engine. / They have low boiling points and may vaporise when the engine gets hot *[1 mark for each correct answer]*.

Pages 59-61: Environmental Problems

1 a) released *[1 mark]*, combustion *[1 mark]*, oxidised *[1 mark]*

b) hydrocarbon + **oxygen** → **water** + carbon dioxide *[1 mark for each answer]*

c) i) The percentage of carbon dioxide in the atmosphere is increasing at an increasing (exponential) rate *[1 mark]*.

ii) The increase in carbon dioxide in the atmosphere is causing climate change/the average temperature to increase/global warming *[1 mark]*.

d) i) Any two from: e.g. biofuels are renewable/they won't run out. / Biofuels are carbon neutral/won't contribute to climate change. / Fossil fuels often contain sulfur and release sulfur dioxide which causes acid rain, whereas biofuels don't. *[1 mark for each advantage]*.

ii) Any two from: e.g. farmers might start growing crops for biofuel production instead of growing food. This could cause an increase in food prices or even famine. / We can't make enough biodiesel to replace the amount of diesel we currently use. / Ethanol isn't currently widely available. / Engines need to be converted before they'll work with ethanol fuels *[1 mark for each disadvantage]*.

2 a) In hydrogen fuel cells *[1 mark]*.

b) Burning hydrocarbon fuels produces lots of pollutants. Burning hydrogen only produces water *[1 mark]*.

c) Any two from: e.g. you need a special expensive engine to burn hydrogen./Ordinary cars would need to be converted to burn hydrogen. / Hydrogen is hard to store/transport (because it takes up lots of space and is explosive). / Hydrogen isn't currently a widely available fuel./Fuelling stations would need to be converted in order to supply hydrogen. / Energy from another source is needed to produce hydrogen. This usually comes from burning fossil fuels, which wipes out the environmental benefits *[1 mark for each reason]*.

3 a) hydrogen *[1 mark]*, carbon *[1 mark]*

b) Plant materials are fermented *[1 mark]* at a temperature between 20 and 35 °C *[1 mark]*. This produces a dilute solution of ethanol, from which pure ethanol can be separated *[1 mark]*.

c) i) biofuel *[1 mark]*

ii) e.g. biodiesel *[1 mark]*

4 a) i) carbon/soot *[1 mark]*, unburnt fuel *[1 mark]*

ii) Particles form if there is not enough oxygen for the fuel to burn completely *[1 mark]*, and there will be less oxygen available in an internal combustion engine *[1 mark]*.

iii) global dimming *[1 mark]*

b) i) At high temperatures *[1 mark]*.

ii) E.g. acid rain can acidify lakes / cause wildlife to die / damage or kill trees / damage limestone buildings *[1 mark]*.

Any sensible effect of acid rain will get you the mark here.

c) i) acid rain *[1 mark]*

ii) E.g. power stations *[1 mark]*, e.g. cars/vehicles/diesel vehicles *[1 mark]*

Page 62: Cracking Crude Oil

1 a) i) The breakdown of long hydrocarbons *[1 mark]* to form short-chain molecules, which are more useful *[1 mark]*.

ii) alkanes *[1 mark]*, alkenes *[1 mark]*

iii) E.g. alkanes are useful as fuels (like petrol and paraffin) / Alkenes are useful for making plastics (like poly(ethene)) *[1 mark]*.

b) i) It acts as a catalyst / catalyses the reaction *[1 mark]*.

ii) alkenes *[1 mark]*, (small) alkanes *[1 mark]*

iii) E.g. heat from the Bunsen burner — place a heat-proof mat underneath the Bunsen burner/wear heat-proof gloves. / If the boiling tube cools, water could be sucked back and shatter it — disconnect the delivery tube before reducing the heat. / Paraffin fumes could irritate the eyes — wear safety goggles. / Hydrocarbons are highly flammable — make sure that the paraffin and the products of the reaction are kept well away from the Bunsen burner flame. *[1 mark for any sensible hazard and 1 mark for a relevant precaution]*.

c) Mixing with steam and heating to a very high temperature (300 °C) *[1 mark]*.

d) thermal decomposition *[1 mark]*

In cracking, you break down molecules by heating them — and that's thermal decomposition.

Pages 63-64: Alkenes and Ethanol

1 a) C_nH_{2n} *[1 mark]*

b)

Name of alkene	Formula	Displayed formula
Ethene	C_2H_4	H\C=C/H with H H
Propene	C_3H_6	H\C=C—C—H with H, H, H, H H

[4 marks available — 1 mark for each correct name or formula.]

The displayed formula for propene could also be drawn with the double bond between the second and third carbon atoms. As long as you've got the right number of hydrogen atoms attached to each carbon atom, it doesn't matter which two carbons the double bond is between.

2 a) i)   C and D *[1 mark]*

   ii)   The molecules have a double covalent bond between two of the carbon atoms in their chain *[1 mark]*.

   b)   When bromine water was added to compound C and shaken the solution changed from orange to colourless *[1 mark]*. When bromine water was added to compound E and shaken the solution remained orange *[1 mark]*.

3 a) H_2O *[1 mark]*

   b) i)   ethene *[1 mark]*

   ii)   alkenes *[1 mark]*

   c)   a catalyst *[1 mark]*

   d)   Method A. Crude oil is the raw material from which ethene is made, and there is a good supply of crude oil in Country Z./In Country Z the workers are paid high wages. This means that method B, which needs lots of labour, would be very expensive to use *[1 mark]*.

4   36 °C – 18 °C *[1 mark]*
   = **18 °C** *[1 mark]*
   [2 marks for a correct final answer, otherwise 1 mark for a final answer calculated correctly from a misread scale.]

Pages 65-66: Polymers

1 a) i)

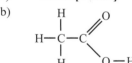

   [1 mark for the structure of the polymer, 1 mark for the bonds continuing either side of the structure, 1 mark for putting brackets around the structure and n after it]

   ii)   Joining lots of small alkene molecules/monomers to form very large molecules (polymers) *[1 mark]*.

   b) i)   Polymer 1/LDP, because freezer bags need to be flexible *[1 mark]*.

   ii)   A *[1 mark]*

   iii)   E.g. by using different catalysts/reaction conditions for the polymerisation reaction *[1 mark]*.

2 a)   They aren't biodegradable/are not broken down by microorganisms *[1 mark]*. So they do not break down in landfill sites *[1 mark]*.

Plastics also create a huge litter problem — once they're out in the environment, they don't break down.

   b)   cornstarch *[1 mark]*

   c)   The polymer chains in thermosoftening plastics are held together by weak intermolecular forces *[1 mark]*. This means they have low melting points *[1 mark]*. In thermosetting plastics the polymer chains are held together by covalent bonds or cross-linking bridges *[1 mark]*. This means they have higher melting points *[1 mark]*.

3 a)   As the temperature rises from 10 to 25 °C, the polymer needs much less pressure to squash it (it gets easier to squash) *[1 mark]*. As the temperature rises above 25 °C, the amount of pressure needed to squash the polymer stays about the same (the polymer doesn't get much easier to squash) *[1 mark]*.

   b)   The part of the pillow under the person's head will get warmed and become softer/mould to the shape of the user's head *[1 mark]*.

This is how memory foam mattresses and pillow work.

Page 67: Alcohols

1 a)

Name of alcohol	Number of carbon atoms	Molecular formula	Displayed formula
Methanol	1	CH_3OH	H—C—O—H with H, H
Ethanol	2	C_2H_5OH	H—C—C—O—H with H H, H H
Propanol	3	C_3H_7OH	H—C—C—C—O—H with H H H, H H

   [6 marks available — 1 mark for each correct name or formula.]

   b)   -OH *[1 mark]*

   c) i)   7 *[1 mark]*

   ii)   Ethanol *[1 mark]*, e.g. solvents *[1 mark]*, e.g. fuels *[1 mark]*

It doesn't matter which way round you write the last two answers here, as long as they're both correct uses of alcohols.

   d)   carbon dioxide *[1 mark]*, water *[1 mark]*

   e)   butanol *[1 mark]*, $C_5H_{11}OH$ *[1 mark]*

   f) $2C_3H_7OH + 2Na \rightarrow 2C_3H_7ONa + H_2$
   [1 mark]

Page 68: Carboxylic Acids

1 a)   -COOH *[1 mark]*

   b)   H—C—C with O and O—H, H at bottom
   [1 mark]

   c)   ethanol *[1 mark]*, microbes *[1 mark]*, vinegar *[1 mark]*

   d) i)   Ethanoic acid ionises/releases H^+ ions when it is dissolved in water, making the solution acidic *[1 mark]*. But it doesn't ionise completely so it is a weak acid *[1 mark]*.

   ii)   You would expect the pH of the hydrochloric acid to be lower *[1 mark]*.

2 a)   an ester *[1 mark]*

   b) $2CH_3CH_2COOH + K_2CO_3 \rightarrow$
   $CO_2 + H_2O + 2CH_3CH_2COOK$
   [1 mark for all reactants and products correct, 1 mark for balancing]

Page 69: Esters

1 a)   E.g. sulfuric acid *[1 mark]*

It's usually sulfuric acid that's used as the catalyst for this reaction, but any named acid would be fine here.

   b) $CH_3COOH + CH_3CH_2OH \rightarrow CH_3COOCH_2CH_3 + H_2O$
   [1 mark]

   c)   ethyl ethanoate *[1 mark]*

   d)   -COO-

137

e) E.g. the students could use a water bath to heat the reaction mixture. / The students could use a Bunsen burner to heat some water in a beaker and place a test tube containing the reaction mixture into the hot water. / The students could heat the reaction mixture using an electric heating mantle *[1 mark]*.

You can have the mark here for any sensible method of heating the reaction mixture that doesn't involve holding it directly over an open flame.

2 a) A volatile chemical evaporates easily *[1 mark]*.
   b) They often smell nice/have distinctive smells *[1 mark]*.
   c) E.g. as flavourings / as solvents *[1 mark]*.

Section Seven — Energy and Equilibria

Page 70: Energy Transfer in Reactions
1   endothermic *[1 mark]*, exothermic *[1 mark]*, greater *[1 mark]*
2 a) Endothermic. The temperature decrease shows that the reaction is taking in energy from the surroundings *[1 mark]*.
   b) A–B. The reaction is endothermic, so more heat energy must be taken in when this bond is broken than is released when the A–C bond is formed *[1 mark]*.
   c) i) The overall change in energy during the reaction *[1 mark]*.
      ii) positive *[1 mark]*
      iii) ΔH *[1 mark]*

Page 71: Energy Transfers and Reversible Reactions
1 a) E.g. to make a self-heating can / in a hand warmer *[1 mark]*.
You can have any sensible everyday use for an exothermic reaction here...
   b) E.g. in a cooling pack for sports injuries *[1 mark]*
...and any sensible everyday use for an endothermic reaction here.
2 a) thermal decomposition *[1 mark]*
   b) Endothermic, because it takes in heat energy *[1 mark]*.
3 a) +78.2 kJ/mol *[1 mark]*.
   b) i) It turns from white to blue *[1 mark]*.
      ii) It increases *[1 mark]*.

Page 72: Energy Level Diagrams
1 a) The minimum energy needed by reacting particles for a reaction to occur *[1 mark]*.
   b) They provide a different pathway with a lower activation energy *[1 mark]*.
   c) i) A / C *[1 mark]*
      ii) B / D *[1 mark]*
      iii) C *[1 mark]*
      iv) B *[1 mark]*
   d) i) –90 kJ/mol *[1 mark]*
      ii) E.g.

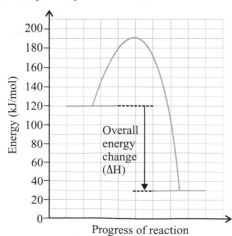

[1 mark for adding a downward arrow to the graph between 120 kJ/mol and 30 kJ/mol and labelling it as the overall energy change]
      iii) 70 kJ/mol *[1 mark]*

iv) E.g.

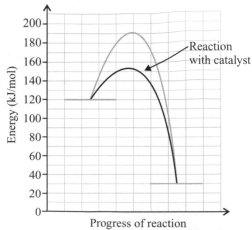

[1 mark for a curve starting and finishing at the same energy level as the original curve, but having a lower peak/activation energy]

Page 73: Bond Dissociation Energy
1   $(4 \times 412) + (2 \times 498) = 2644$ kJ/mol *[1 mark]*
    $(2 \times 743) + (4 \times 463) = 3338$ kJ/mol *[1 mark]*
    Enthalpy change = $2644 - 3338 = $ **–694 kJ/mol** *[1 mark]*
2   $158 + (4 \times 391) + 498 = 2220$ kJ/mol *[1 mark]*
    $945 + (4 \times 463) = 2797$ kJ/mol *[1 mark]*
    Enthalpy change = $2220 - 2797 = $ **–577 kJ/mol** *[1 mark]*
Don't forget to double-check the sign of your answer — if it's wrong, you'll lose the mark.

Pages 74-75: Measuring Energy Transfer
1 a) i) E.g. to insulate the cup *[1 mark]*.
      ii) E.g. to reduce energy/heat lost through evaporation *[1 mark]*.
   b) 31 °C – 21 °C = **10 °C** *[1 mark]*
   c) i) Some energy is always lost to the surroundings *[1 mark]*.
      ii) The experiment could be repeated and an average temperature change calculated *[1 mark]*.
   d) E.g. neutralisation *[1 mark]*.
2 a) Because copper conducts heat very well *[1 mark]*.
   b) Heat energy change = $50 \times 4.2 \times 30.5 = $ **6405 J** *[1 mark]*
   c) Energy produced = $6405 \div 0.7$ *[1 mark]*
      = 9150 J/g = **9.15 kJ/g** *[1 mark]*
3 a) M_r of ethanol = $(2 \times 12) + (6 \times 1) + (1 \times 16) = 46$ *[1 mark]*
      Number of moles = $1.15 \div 46 = $ **0.025** *[1 mark]*
   b) Molar enthalpy change = $-7.245 \div 0.025$ *[1 mark]*
      = **–289.8 kJ/mol** *[1 mark]*
The enthalpy change (–7.245 kJ) is negative because combustion is an exothermic reaction.

Pages 76-77: Equilibrium and Yield
1 a) It is reversible *[1 mark]*.
   b) i) It takes in heat *[1 mark]*, because it's endothermic/ all reversible reactions are exothermic in one direction and endothermic in the other direction. *[1 mark]*
      ii) The forward reaction, as lowering the temperature favours the exothermic reaction *[1 mark]*.
      iii) One reaction is always exothermic and the other reaction endothermic *[1 mark]*, so a change in temperature will always favour one reaction more than the other *[1 mark]*.
   c) It won't affect the position of equilibrium *[1 mark]*, because there are the same number of gas molecules on either side of the equation *[1 mark]*.
   d) It is not a closed system / the products can escape *[1 mark]*.
Remember — you only get an equilibrium when a reversible reaction takes place in a closed system.
2 a) calcium carbonate/$CaCO_3$ *[1 mark]*
   b) the same/an equal *[1 mark]*

3   The first student's reaction conditions are better *[1 mark]*. The forward reaction is endothermic *[1 mark]*, so using a high temperature will increase this reaction to use up the extra heat *[1 mark]*. There are more gas molecules/moles on the right-hand side of the equation *[1 mark]*, and a lower pressure will encourage the reaction that produces more molecules/moles of gas *[1 mark]*.

You can tell that the forward reaction is endothermic because the reaction mixture needs to be heated for the forward reaction to happen.

Page 78: The Haber Process

1 a) $N_2 + 3H_2 \rightleftharpoons 2NH_3$
[1 mark for correct reactants, 1 mark for correct balancing]

b)   Reactant 1: nitrogen *[1 mark]* Source: the air *[1 mark]*
Reactant 2: hydrogen *[1 mark]* Source: natural gas/cracking of hydrocarbons/electrolysis of water *[1 mark]*

It doesn't matter which way round you have the reactants — each one just needs to be matched up with a correct source.

2 a) i)   200 atmospheres *[1 mark]*

ii)   Because the forward reaction is exothermic, using a high temperature shifts the equilibrium to the left *[1 mark]* and decreases the yield of ammonia *[1 mark]*. However, the reaction happens faster at higher temperatures *[1 mark]*. The temperature used is a compromise between high yield and a fast reaction *[1 mark]*.

b)   iron *[1 mark]*

c)   E.g. as the ammonia gas cools it condenses and liquefies *[1 mark]*, which means it separates out from the other gases and can be easily piped off *[1 mark]*. The unused hydrogen and nitrogen are recycled/re-circulated/pumped back into the reaction chamber to be used in the reaction again *[1 mark]*.

Section Eight — Electrolysis and Analysis

Page 79: Electrolysis

1 a)   (molten) lead/Pb *[1 mark]*

b)   For electricity to flow through the electrolyte the ions need to be free to move *[1 mark]*. In solid ionic compounds the ions are in fixed positions *[1 mark]*, but when they are molten or dissolved the ions can move about *[1 mark]*.

c) i) $Pb^{2+} + 2e^- \rightarrow Pb$
[1 mark for 2e⁻]
$2Br^- \rightarrow Br_2 + 2e^-$
[1 mark for 2Br⁻, 1 mark for 2e⁻]

ii)   At the cathode/negative electrode, because reduction is the gain of electrons *[1 mark]* and at the cathode/negative electrode the lead ions/positive ions gain electrons *[1 mark]*.

iii)   The bromide ions are negatively charged *[1 mark]*, so they are attracted towards the positively charged anode *[1 mark]*.

Page 80: Electrolysis of Sodium Chloride Solution

1 a) $2Cl^- \rightarrow Cl_2 + 2e^-$
[1 mark for correct reactants and products, 1 mark for correct balancing]

You can also have the marks if you showed the two electrons being subtracted from the left hand side of the equation instead.
Like this: $2Cl^- - 2e^- \rightarrow Cl_2$

b)   hydrogen (gas)/H_2 *[1 mark]*

c)   The sodium ions stayed in solution (because they're more reactive than hydrogen). *[1 mark]*. Hydroxide ions from water were also left behind *[1 mark]*. This means that sodium hydroxide (NaOH) was left in the solution *[1 mark]*.

d) i)   Chloride ions are more reactive than hydroxide ions *[1 mark]*, so with a dilute solution, the chloride ions stay in solution and oxygen is formed (from the hydroxide ions) *[1 mark]*. But in a concentrated solution, there are many more chloride ions present than hydroxide ions *[1 mark]*, so chlorine is formed instead *[1 mark]*.

ii)   hydrogen (gas)/H_2 *[1 mark]*

e) i)   e.g. soap *[1 mark]*

ii)   e.g. bleach/plastics *[1 mark]*

Page 81: Electrolysis of Aluminium and Electroplating

1 a)   The cost of electricity / a lot of electricity/energy is needed *[1 mark]*.

b) i)   It reduces the temperature needed to melt the electrolyte/aluminium oxide *[1 mark]*, meaning that less energy/electricity is used *[1 mark]*.

ii) $Al^{3+} + 3e^- \rightarrow Al$
$2O^{2-} \rightarrow O_2 + 4e^-$
[1 mark for correct reactants and product in the aluminium equation, 1 mark for correct reactant and products in the oxygen equation, 1 mark for balancing both half-equations correctly]

iii)   The positive electrode is made from carbon *[1 mark]*. The oxygen produced at the positive electrode reacts with the carbon (to form carbon dioxide) *[1 mark]*.

2 a)   silver ions/Ag^+ *[1 mark]*

b)   E.g. to make it look as attractive as a silver fork without the expense of making it from solid silver / to make it more resistant to corrosion *[1 mark]*.

c)   E.g. to make it more durable / to make it resistant to corrosion / to make it look more attractive / to allow it to conduct electricity (more effectively)
[1 mark for any correct answer that is different to the reason given in part b)].

Pages 82-84: Tests for Positive and Negative Ions

1 a)   E.g. take a clean platinum wire loop, dip it into the substance to be tested and put the material into a blue Bunsen burner flame/the hot part of a Bunsen burner flame.
[1 mark for any suitable method stated to transfer the material into the flame, 1 mark for saying that the material needs to be placed in the flame]

b)   Potassium would give a lilac flame *[1 mark]* but sodium would give a yellow flame *[1 mark]*.

c)   The medicines also contain sodium ions *[1 mark]*, so the colour produced by the sodium could interfere with the flame test result *[1 mark]*.

2) a)

Metal ion	Colour of precipitate
Fe^{2+}	green
Cu^{2+}	blue
Fe^{3+}	brown

[1 mark for each ion or colour]

b) Fe^{2+} (aq) + $2OH^-$ (aq) $\rightarrow$ $Fe(OH)_2$ (s)
[1 mark for all state symbols correct]

3 a) i) SO_4^{2-} *[1 mark]*

ii) I^- *[1 mark]*

iii) CO_3^{2-} *[1 mark]*

b) i)   dilute hydrochloric acid *[1 mark]*, barium chloride solution *[1 mark]*

ii)   a white precipitate *[1 mark]*

c) i)   e.g. dilute hydrochloric acid *[1 mark]*

Any dilute acid will do here — you'd usually use hydrochloric acid, but sulfuric acid or nitric acid would work too.

ii)   carbon dioxide *[1 mark]*

d)   Add dilute nitric acid *[1 mark]* followed by silver nitrate solution *[1 mark]*. Chloride ions will give a white precipitate *[1 mark]* whereas bromide ions will give a cream precipitate *[1 mark]*.

e) i) Cl^- *[1 mark]*

ii) SO_4^{2-} *[1 mark]*

4 a) i) The formation of a white precipitate could also indicate the presence of calcium ions or magnesium ions *[1 mark]*.
   ii) The compound does not contain aluminium ions *[1 mark]*. If the compound contained aluminium ions, the precipitate would have dissolved in the excess sodium hydroxide *[1 mark]*.
 b) i) E.g. error: The student used a yellow Bunsen flame. Correction: Use a blue Bunsen flame / Error: The student cleaned the wire loop with paper towel. Correction: Dip the wire loop in hydrochloric acid and rinse with distilled water. *[1 mark for a sensible error, 1 mark for a sensible correction to that error]*
   ii) E.g. wear eye protection / wear gloves when handling chemicals / stand the Bunsen burner on a heat-proof mat. *[1 mark for any sensible safety precaution]*
   iii) red *[1 mark]*
 c) calcium iodide *[1 mark]*

Page 85: Separating Mixtures

1 a) elements *[1 mark]*, physical *[1 mark]*
 b)

Temperature on thermometer	Contents of the beaker
30 °C	no liquid
65 °C	propanone
110 °C	both liquids

   [2 marks for all three rows correct, otherwise 1 mark for any two rows correct]

2 a) E.g. mix the lawn sand with water to dissolve the ammonium sulfate *[1 mark]*. Filter the mixture using filter paper to remove the sharp sand *[1 mark]*. Wash the sand to remove any ammonium sulfate *[1 mark]*. Pour the remaining solution into an evaporating dish and slowly heat it/leave it for a while to evaporate most of the water (and form crystals) *[1 mark]*. Dry the products in a drying oven/desiccator *[1 mark]*.
 b) E.g. the products were not completely dry *[1 mark]*.
When you're crystallising a salt, if you don't dry the crystals off properly, you'll end up including the mass of some water when you weigh your product. (This can also throw your calculations off if you want to find the percentage yield of the reaction.)

Page 86: Paper Chromatography

1 a) E.g. draw a pencil line near the bottom of a sheet of filter paper and add spots of different inks to the line at intervals. *[1 mark]*. Put the paper in a beaker of a solvent *[1 mark]*, so that the pencil line and the spots of ink are above the level of the solvent *[1 mark]*.
 b) Printers A and C. *[1 mark for both]*
 c) The chemicals have different solubilities in the solvent/ some of the chemicals dissolve more readily in the solvent than others *[1 mark]*, so the chemicals travel up the paper at different rates *[1 mark]*.
 d) i) Distance travelled by solvent: 6.0 cm / 60 mm
   Distance travelled by chemical Y: 3.3 cm / 33 mm
   [1 mark for both correct, 1 mark for correct units]
Your measurements can be ± 0.1 cm if you measured in centimetres, or ± 1 mm if you measured in millimetres. To get the second mark, you must give units and they must match your measurements.
   ii) $3.3 / 6.0 = $ **0.55** *[1 mark]*
If you got part d) i) wrong, you can still have the mark in part d) ii) for putting your numbers into the R_f formula correctly.

Page 87: Instrumental Methods

1 a) E.g. they're very fast / they're very accurate / they're very sensitive/can detect even the tiniest amounts of a substance *[1 mark for each correct advantage]*.
 b) How to grade your answer:
   0 marks: No relevant points are made.
   1-2 marks: A brief description of the process is given.
   3-4 marks: A good description of the process is given, but some detail is missing. The answer has a logical structure and spelling, grammar and punctuation are mostly correct.
   5-6 marks: A full and detailed description of the process is given. The answer has a logical structure and uses correct spelling, grammar and punctuation.
Here are some points your answer may include:
A gas is used to carry a sample of the substance through a column packed with a solid material.
The compounds in the substance travel through the column at different speeds, so they separate out.
A recorder draws a gas chromatograph. The number of peaks shows the number of different compounds that are present in the sample.
The position of the peaks shows the retention time of each compound.
The output from the gas chromatograph is fed into the mass spectrometer in order to identify the separated compounds.
The mass spectrometer produces a graph which shows the molecular ion peak of each compound in the sample.
You can use this to work out the relative molecular mass of each compound.
The relative molecular mass can be used to identify the compound.

The Periodic Table

1
H
Hydrogen
1

Atomic number ↗

Periods

	Group 1	Group 2												Group 3	Group 4	Group 5	Group 6	Group 7	Group 0
1																			4 **He** Helium 2
2	7 **Li** Lithium 3	9 **Be** Beryllium 4												11 **B** Boron 5	12 **C** Carbon 6	14 **N** Nitrogen 7	16 **O** Oxygen 8	19 **F** Fluorine 9	20 **Ne** Neon 10
3	23 **Na** Sodium 11	24 **Mg** Magnesium 12												27 **Al** Aluminium 13	28 **Si** Silicon 14	31 **P** Phosphorus 15	32 **S** Sulfur 16	35.5 **Cl** Chlorine 17	40 **Ar** Argon 18
4	39 **K** Potassium 19	40 **Ca** Calcium 20	45 **Sc** Scandium 21	48 **Ti** Titanium 22	51 **V** Vanadium 23	52 **Cr** Chromium 24	55 **Mn** Manganese 25	56 **Fe** Iron 26	59 **Co** Cobalt 27	59 **Ni** Nickel 28	63.5 **Cu** Copper 29	65 **Zn** Zinc 30		70 **Ga** Gallium 31	73 **Ge** Germanium 32	75 **As** Arsenic 33	79 **Se** Selenium 34	80 **Br** Bromine 35	84 **Kr** Krypton 36
5	85 **Rb** Rubidium 37	88 **Sr** Strontium 38	89 **Y** Yttrium 39	91 **Zr** Zirconium 40	93 **Nb** Niobium 41	96 **Mo** Molybdenum 42	98 **Tc** Technetium 43	101 **Ru** Ruthenium 44	103 **Rh** Rhodium 45	106 **Pd** Palladium 46	108 **Ag** Silver 47	112 **Cd** Cadmium 48		115 **In** Indium 49	119 **Sn** Tin 50	122 **Sb** Antimony 51	128 **Te** Tellurium 52	127 **I** Iodine 53	131 **Xe** Xenon 54
6	133 **Cs** Caesium 55	137 **Ba** Barium 56	139 **La** Lanthanum 57	178 **Hf** Hafnium 72	181 **Ta** Tantalum 73	184 **W** Tungsten 74	186 **Re** Rhenium 75	190 **Os** Osmium 76	192 **Ir** Iridium 77	195 **Pt** Platinum 78	197 **Au** Gold 79	201 **Hg** Mercury 80		204 **Tl** Thallium 81	207 **Pb** Lead 82	209 **Bi** Bismuth 83	209 **Po** Polonium 84	210 **At** Astatine 85	222 **Rn** Radon 86
7	223 **Fr** Francium 87	226 **Ra** Radium 88	227 **Ac** Actinium 89	261 **Rf** Rutherfordium 104	262 **Db** Dubnium 105	266 **Sg** Seaborgium 106	264 **Bh** Bohrium 107	277 **Hs** Hassium 108	268 **Mt** Meitnerium 109	271 **Ds** Darmstadtium 110	272 **Rg** Roentgenium 111								